Principles of Water Resources: Development and Management

Principles of Water Resources: Development and Management

Martin Green

MURPHY & MOORE
www.murphy-moorepublishing.com

Principles of Water Resources: Development and Management
Martin Green
ISBN: 978-1-63987-459-0 (Hardback)

MURPHY & MOORE

Published by Murphy & Moore Publishing,
1 Rockefeller Plaza,
New York City, NY 10020, USA

Cataloging-in-Publication Data

Principles of water resources : development and management / Martin Green.
p. cm.
Includes bibliographical references and index.
ISBN 978-1-63987-459-0
1. Water resources development. 2. Water-supply--Management.
3. Water-supply--Environmental aspects. I. Green, Martin.
TC405 .P75 2021
333.91--dc23

For more information regarding Murphy & Moore Publishing and its products, please visit the publisher's website www.murphy-moorepublishing.com

Table of Contents

Preface

The natural sources of water which can be used for various industrial, agricultural, recreational and household activities are termed as water resources. Surface water, under river flows, ground water and frozen water fall under this category. Water resource management refers to those activities which aim to develop, plan, manage and distribute the water resources in the most efficient manner. Some of the commonly used techniques for reclaiming freshwater are desalination, groundwater recharging and reverse osmosis. The key objective of water resource management is to consider all the demands of water, and allocate it on an equitable basis. The sustainability of current as well as future water resources is one of the biggest challenges of this field. This book is a compilation of chapters that discuss the most vital concepts related to the development and management of water resources. Some of the diverse topics covered herein address the varied branches that fall under this category. This book will provide comprehensive knowledge to the readers.

A detailed account of the significant topics covered in this book is provided below:

Chapter 1- Water is an inorganic chemical substance which is transparent, odorless, tasteless and colorless, and is the main constituent of fluids of all living organisms and hydrosphere. This is an introductory chapter which will briefly introduce all the significant aspects of water such as water resources, hydrologic cycle, groundwater hydrology, water balance, etc.

Chapter 2- There are many types of modeling approaches to provide a better understanding of alternative water resource designs and management plans. Some of these modeling approaches are catchment modeling, river basin modeling and rainfall-runoff modeling. All these types of modeling in water resources have been carefully analyzed in this chapter.

Chapter 3- Water conservation refers to the strategies and activities required to reduce unnecessary water usage to protect the hydrosphere. A few of the aspects related to water conservation are water harvesting, integrated water resources management and integrated urban water management. This chapter has been carefully written to provide an easy understanding of the varied facets of water conservation.

Chapter 4- The contamination of water bodies by harmful substances as a result of human activities is known as water pollution. Some of the topics studied under water pollution are groundwater pollution, flow dynamics and wastewater. The topics elaborated in this chapter will help in gaining a better perspective about water pollution.

Chapter 5- Irrigation is an artificial process of supplying water to land or crops to assist in the production of crops. A few of the strategies studied in relation to irrigation are deficit irrigation, flexible barge, hippo water roller and peak water. This chapter closely examines the key concepts of irrigation to provide an extensive understanding of the subject.

It gives me an immense pleasure to thank our entire team for their efforts. Finally in the end, I would like to thank my family and colleagues who have been a great source of inspiration and support.

Martin Green

1

Water Resources: An Introduction

Water is an inorganic chemical substance which is transparent, odorless, tasteless and colorless, and is the main constituent of fluids of all living organisms and hydrosphere. This is an introductory chapter which will briefly introduce all the significant aspects of water such as water resources, hydrologic cycle, groundwater hydrology, water balance, etc.

Water

Water is necessity for survival of life on earth. About 70% of Earth (often called as Blue Planet) is covered with water and as far as we know, it is the only planet in our solar system containing liquid water. The percentage of water in human body having 70 kg of weight is 67.85%. Many philosophers, scientists, saints from different sects including holy books quote the importance of water. Some of them like: "O water, because of your presence, the atmosphere is so refreshing, and imparts us with vigour and strength. We respect you who gladden us by your pure essence."

The role of water in evolution and philosophical context can be understood as: "Although our ancestors have been terrestrial for so long, each person spends the first nine months of life as aquatic as fish. Within the womb the fetus is buoyed up by a warm sea that is almost identical with bold plasma, and hence brackish-a third as rich in salts as the oceans of the earth."

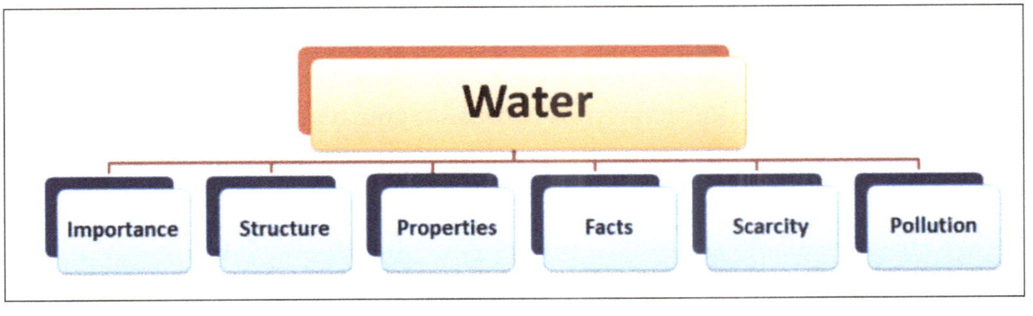

There are different terms and definitions related to water such as drinking water which can be defined as water for ingestion, basic personal and domestic hygiene and cooking. It excludes water for clothes washing, an activity that frequently happens at the water source, water point, in rivers or streams. Whereas, safe drinking water means that the water meets does not represent any significant risk to health over a lifetime of consumption, including different sensitivities that may occur between life stages.

It should be colourless, odourless, pleasant taste and free from toxic chemicals and disease causing microorganisms. The United Nations Millennium Development Goals (MDG) Target 7C called on countries to "Halve, by 2015, the proportion of population without sustainable access to safe drinking-water and basic sanitation". An improved drinking water source is defined as a type of drinking water facility or water delivery point that by the nature of its design protects the drinking water source from external contamination, particularly of fecal origin. Piped water into dwelling, plot or yard, public tap/standpipe, tubewell/borehole, protected dug well, protected spring and rainwater are included under improved whereas unprotected dug well and spring, tanker truck, small cart with tank/drum, and surface water (river, dam, lake, pond, stream, channel, irrigation channel) are considered as unimproved drinking water sources. Over 90% of the world's population now has access to improved sources of drinking water.

Accessibility to water as defined in the "Right to Water" includes a continuous supply of a minimum amount of water sufficient for drinking, personal and domestic hygiene, for an affordable price, within a reasonable distance. According to WHO basic access can be defined as the availability of at least 20 litres of drinking water per person per day within a distance of not more than one km of the dwelling, corresponding to a maximum water hauling round trip of 30 minutes. This definition is more appropriate for rural areas as distance to a source is not a problem in urban areas. The United Nations General Assembly on 28 July 2010, through Resolution 64/292 unambiguously recognized the human right to water and sanitation and acknowledged that clean drinking water and sanitation are essential to the realization of all human rights. The resolution calls upon states and international organizations to provide financial resources, help capacity-building and technology transfer to help countries, in particular developing countries, to provide safe, clean, accessible and affordable drinking water and sanitation for all. The study of water is known as hydrology and is divided into a number of subcategories. Limnology is the branch of the science dealing with the characteristics of fresh water including biological, chemical and physical properties, whereas oceanography is the science of ocean including its various characteristics.

There are three available sources of water. First the groundwater from springs and wells which people mainly used for drinking in the first half of the 20th century. Fresh surface water is the second source of drinking water supply. As this water contains plenty of biological life from lower to higher organisms, including fish and frogs therefore considered as "live water" in contrast to "dead water", that is, tap water treated with chlorine, or distilled water. However, due to industrialization, agricultural complexes, methods of mass transport, utilities, communal infrastructures; the growth of cities and settlements all resulted in the mass pollution of surface water. Due to pollution load, the self-purification capacity of water bodies by the natural process is either reduced or lost. The third source of drinking water supply can be seas and oceans, where the average salinity of about 30 g/l (30,000 ppm) makes it unfit for direct use. Through main desalination technologies of distillation, membrane technology, and electrochemical the sea water is desalinized. The desalinated water is not drinking water. It is necessary to make adjustments to modify the salt composition. Such waters should be "conditioned". This could be one of the best options where there is no alternative source of fresh water.

Structure of Water Molecule

Water's properties can best be understood by considering the structure and bonding of the water molecule. The water molecule is made up of two hydrogen atoms bonded to an oxygen atom. The bonds which hold the hydrogen and oxygen together are called covalent bonds and are very strong. The three atoms are not in a straight line; instead, they form an angle of 104.5°. The water molecule maintains a bent shape (bent at 107.5° actually) because of two reasons. First the tetrahedral arrangement around the oxygen and second the presence of lone pair electrons on the oxygen. These are the electrons that are not involved in the covalent bonds and are left alone. These lone pairs are very negative - containing two negative electrons each and want to stay away from each other as much as possible. These repulsive forces act to push the hydrogens closer together. The importance of the bent structure of water is that it provides water with two distinct "sides": one side of the water molecule has two negative lone pairs, while the other side presents two hydrogens.

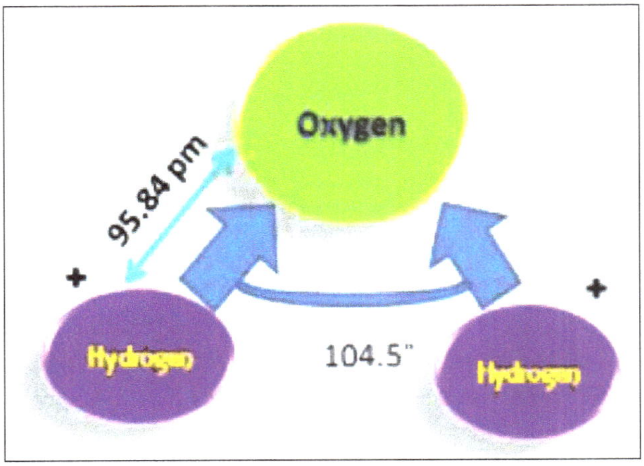

Structure of water molecule.

The hydrogen's are slightly positive because of the electro-negativity of oxygen. Electro-negativity is a measure of how much one atom wants to have electrons, and oxygen wants to have electrons more than hydrogen does. Oxygen has a higher electro-negativity. Due to the difference in electronegativity, the electrons in the covalent bonds between oxygen and hydrogen get pulled slightly toward the oxygen. This leaves the hydrogen's a little bit electron-deficient and thus slightly positive. Because water has a slightly negative end and a slightly positive end, it can interact with itself and form a highly organized 'inter-molecular' network.

The positive hydrogen end of one molecule can interact favourably with the negative lone pair of another water molecule. This interaction is call as "Hydrogen bonding" which is a type of weak electrostatic attraction (positive to negative). Because each and every water molecules can form four Hydrogen Bonds, an elaborate network of molecules is formed. The polarity also allows water interacts with an electric field and to interact with other polar molecules - which is how substances get dissolved in water. Because of water's bent structure and the fact that the oxygen atom attracts the negative electrons more strongly than do the hydrogen atoms, the water molecule behaves like a dipole having opposite electrical charges at either end. The water dipole may be attracted to either positively or negatively charged ions.

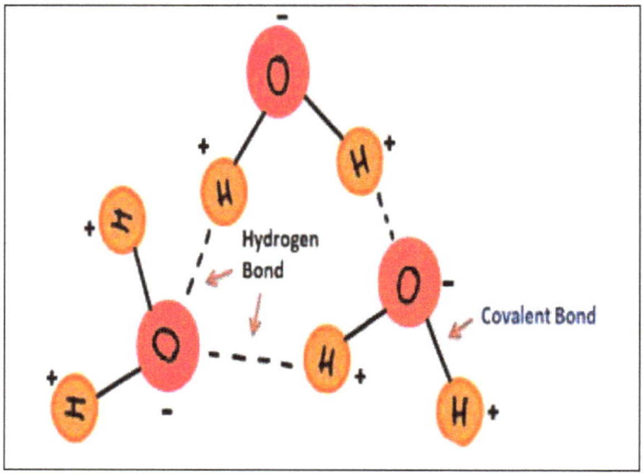

Different bonds in water molecule.

Unique Properties of Water

No other substance on the globe can exist simultaneously in all three phase states i.e. liquid, solid, and vapour. Although water is an apparently simple molecule, it has a highly complex and anomalous character due to its intra-molecular hydrogen bonding. As a gas, water is one of lightest known, as a liquid it is much denser than expected and as a solid it is much lighter than expected when compared with its liquid form. The continual movement of water around the globe is known as the hydrologic cycle. Due to several processes of the cycle i.e. evaporation, transpiration, condensation, precipitation, surface runoff, percolation and collection, water will change from one phase to another.

Liquid water is also characterized by an amazing variety of states from super-cooled to −70 °C, fresh, brines, hydrates, fog, clouds, water bound in cells of an organism, and pseudo liquefied water. Also one must remember that water is an "universal solvent" capable of dissolving many kinds of substances due to which natural water is never absolutely pure. Due to its polar character, water is a good solvent for salts, polar organic compounds and gases. Because of this property almost all biochemical reactions take place in aqueous solutions and water is a key reaction medium for all metabolic processes in living cells. Water dissolves a very large number of molecules in concentrations that are sometimes minute but sufficient to trigger biological or biochemical effects on organisms. As a solvent and a main principal participant of reactions and processes, water is a medium for chemical reactions, biological processes, and various physical phenomena.

Surface Tension

Next to mercury, water has the highest surface tension of all commonly occurring liquids. Water molecules at the surface (next to air) hold closely together, forming an invisible film. The surface tension is denoted by the Greek letter gamma (γ) and is the magnitude F of the force per unit length (L) over which it acts.

$\gamma = F/L$ (The SI unit of surface tension is N/m.)

Surface tension of water at 20 °C and 100 °C are 0.073 and 0.059 N/m, respectively. Water's surface tension can hold weight that would normally sink. Floating a paper piece on top of the water, walking of some aquatic insects e.g. water strider is due to water surface tension property. Surface tension is essential for the transfer of energy from wind to water to create waves. Waves are necessary for rapid oxygen diffusion in lakes and seas.

Capillary Action

Capillary action is defined as the movement of water within the spaces of a porous material due to the forces of adhesion, cohesion, and surface tension. This property of water is essential for the growth of plants and trees where it enters the plant's roots and moves to other parts of the plant through tiny tube-like structures called xylem. By capillary action water reaches more than 90 m above the ground in tallest trees of redwood.

- Adhesion: Water molecules stick to other substances e.g. sponge used to soak-up spilled water.
- Cohesion: Water molecules stick to each other which are due to the hydrogen bonds among the molecules. At the surface water molecules has a much greater attraction for each other than for molecules in the air.

Thermal Properties

Water absorbs or releases more heat than many substances for each degree of temperature increase or decrease. Because of this, it is widely used for cooling and for transferring heat in thermal and chemical processes. Differences in temperature between lakes and rivers and the surrounding air may have a variety of effects. Large bodies of water, such as the oceans or big lakes, have a profound influence on climate. They are the world's great heat reservoirs and heat exchangers and the source of much of the moisture that reaches as rain and snow over adjacent landmasses. This property of water is crucial for stabilizing temperatures on earth.

- Specific Heat: Water has high specific heat of 1.0. The amount of energy required to raise the temperature of water by one degree celsius is quite large. Because so much heat loss or heat input is required to lower or raise the temperature of water, the oceans and other large bodies of water have relatively constant temperatures.
- Heat of Vaporization: Water absorbs heat as it changes from a liquid to a gas; the human body can dissipate excess heat by the evaporation of its sweat. Water's high heat conductivity makes possible the even distribution of heat throughout the body. Water has average Heat of vaporization of about 540 cal/gm or about 0.4 eV/molecule.
- Boiling and Freezing: At sea level pure water boils at 100 °C and freezes at 0 °C, but extra energy is needed to push water molecules into the air. This is called latent heat—the heat required to change water from one phase to another. The high latent heat of evaporation gives resistance to dehydration and considerable evaporative cooling. If a substance is dissolved in water, then the freezing point is lowered. That is why salt is used for street spreading during winter to prevent ice formation. Energy is lost when water freezes. A great amount of heat is released into the environment when liquid water changes to ice. Nights when ice freezes often feel warmer than nights when ice melts.
- Water density: Water is most dense at 4 °C and then begins to expand again (becoming less dense) as the temperature drops further. Liquid water at 4 °C is about 9% denser than ice. This expansion is due to hydrogen bonds in water become more rigid and ordered. As a result, ice floats (frozen water) upon the denser cold water. Water's high surface tension plus its expansion on freezing encourages the erosion of rocks to give soil for our agriculture.
- Solid Expansion: For most substances, solids are denser than liquids. But the special properties of water make it less dense as a solid. This property plays an important role in lake and ocean ecosystems. Floating ice often insulates and protects animals and plants living in the deep water below.
- Specific Heat capacity: It means the resistance to temperature change. It takes a lot of energy to change the water temperature as it takes a lot of energy to break the hydrogen bonds. Water absorbs and stores a large amount of heat from the sun and, as it cools, it releases this heat into the air that moves towards land. During winter season, when the air is colder than the water, it literally picks

up the water's heat while passing over water onto land and warms it. Similarly, during summer when the air is hotter than the water, the air cools off as it travels past the ocean to land and cools the shore. Water's high specific heat capacity also keeps ocean temperatures fairly moderate, which is important for marine life survival.

Isotopes of Water

Hydrogen has two stable isotopes i.e. 1H and 2H (deuterium) and 3H (tritium) a radioactive isotope whereas Oxygen has three stable isotopes of 16O, 17O, and 18O. Distinguished by the mass numbers and characteristics of hydrogen and oxygen, there are eighteen isotopic kinds of water. However, because of the low abundance of the heavier isotopes, almost all water molecules are of three isotopic combinations of $^{1}H_2{}^{16}O$, $^{2}H_2{}^{16}O$ and $^{1}H_2{}^{18}O$. Stable isotopes in water ($^{2}H_2{}^{16}O$ and $^{1}H_2{}^{18}O$) provide the basis for hydrological, paleoclimatic, ecological, archaeological and forensic studies.

Importance of Water in Life

Water is second after oxygen as being vital for life. It is a vital link in the development of mankind, from the earliest times, people settled along rivers, lakes, seas, and oceans. Availability of ample water was one of the main factor behind the establishment of early river-valley civilizations like the Indus Valley Civilization, Ancient Egypt (the Nile), Mesopotamian (along the Tigris and Euphrates rivers), and Chinese civilization (along the Yellow river). People in this world require water for domestic (drinking, cleaning and washing), agricultural (irrigation), industrial, hydro-power generation etc. Humans may feel thirst after a fluid loss of only 1% of bodily fluid and be in danger of death when fluid loss nears 10%. Minimum water requirements for fluid replacement have been estimated at about three liters/day under average temperate climate conditions which can vary according to climate and activity level. According to different organizations and researchers, the average daily water requirement for human survival is between 1.5–5.0 liters per capita per day. Agriculture is the largest consumer of water used by humans worldwide. The standard of living has always been determined to a great extent by the quality of water available. The quality of fresh drinking water depends on the purification level of wastewater.

Use of Water

Water is used for many different purposes throughout the world for domestic (drinking, cooking, personal cleaning, cloth washing, home cleaning, irrigation), industrial and for waste disposal. Globally, industry uses about twice as much water as used in households, mostly for cooling in the production of electricity. Water uses can be classified into consumptive and non-consumptive use. Consumptive use denotes that water which partially or totally used up, through evapotranspiration, transformation, contamination or other processes. This categorization is limited in extent as water never disappears from the water cycle and the amount of water cannot be increased or decreased on the planet. Rather, consumptive uses remove this water from the terrestrial part of the water cycle and return it to a vapour state. Examples of this uses are water for domestic and municipal needs, irrigation and industry. Nearly 85% of total human consumptive use of water worldwide is for irrigation in agriculture. About 70% of the total water used in industry is for cooling purpose only. Nonconsumptive uses do not reduce the volume of water available to that stage of the water cycle and it includes hydropower generation, recreation and water sports, fisheries, inland navigation, and ecosystem maintenance.

Scarcity of Water

Water scarcity involves water stress, water shortage or deficits, and water crisis. The relatively new concept of water stress is the difficulty of obtaining sources of fresh water for use during a period of time and may result in further depletion and deterioration of available water resources. Water scarcity is the lack of sufficient available water resources to meet the demands of water usage within a region. It already affects every continent and around 2.8 billion people around the world at least one month out of every year. More than 1.2 billion people lack access to clean drinking water. Water shortages may be caused by climate change, such as changed weather patterns including droughts or floods, increased pollution and human demands and overuse of water. Water scarcity is being driven by two converging phenomena's of growing freshwater use and depletion of usable freshwater resources.

Water scarcity can be a consequence of two mechanisms where the first physical (absolute) water scarcity is a result of inadequate natural water resources to supply a region's demand, and second economic water scarcity is a result of poor management of the sufficient available water resources. A water crisis is a situation where the available potable, unpolluted water within a region is less than that region's demand. Population growth, industrialization, urbanization and rising affluence in the 20th century resulted in a substantial increase in water consumption. The demand on water resources will continue to increase in coming years. The problem is further aggravated by the uneven water distribution on earth. According to the United Nations Development Programme (UNDP), economic water scarcity is comparatively more contributing in countries or regions experiencing water scarcity, as most countries or regions have enough water to meet domestic, industrial, agricultural, and environmental needs, but lack the means to provide it in an accessible manner. The UN through it MDG Target 7C recognizes the importance of reducing the number of people without sustainable access to clean water and sanitation.

Water Pollution

Water pollution is defined as the addition of any substance to water or changing of water's physical and chemical characteristics in any way which interfere with its use for genuine purposes. In other words we can also define water pollution as excessive concentration of particular substances for sufficient periods of time to cause identifiable effects. The polluted water is unfit for drinking, bathing or washing, agricultural and industrial use, for aquatic life, recreation and aesthetic point of view.

Different types of pathogens, pollutants in water along with poor hygiene practices may result into major illness and mortality. About 3.4 million people die each year from diseases associated with pathogens in water, like cholera, typhoid, infectious hepatitis, polio, cryptosporidiosis, ascariasis and diarrheal diseases. Many of these diseases are due to the presence of human waste in water. Access to quality water is essential for human health and human development. Both are at risk if we fail to stop the pollution. Population growth, increased economic activity, expansion and agriculture intensification, and discharge of untreated sewage into rivers and lakes are the main reasons behind the troubling rise in surface water pollution in many countries including India. Many big drains passing through cities discharge their wastes [including municipal solid wastes (MSW)] directly into rivers.

(a) Nai sota nala (original name Narayani Srotra) showing all types of MSW.

(b) Nai sota nala discharges its waste water along with MSW into river Ganges.

In conclusion water is a vital resource necessary for our survival and proper functioning of the environment. The unique properties and characteristics of it make water use in various activities, which in turn resulted into water pollution, water stress and water scarcity. One should realize the importance of this precious resource and use in such a manner to ensure its sustainability for present as well as for future generations.

Water Resources

Water resources are sources of – usually fresh – water that are useful, or potentially useful, to society; for instance for agricultural, industrial or recreational use. Examples include groundwater, rivers, lakes and reservoirs.

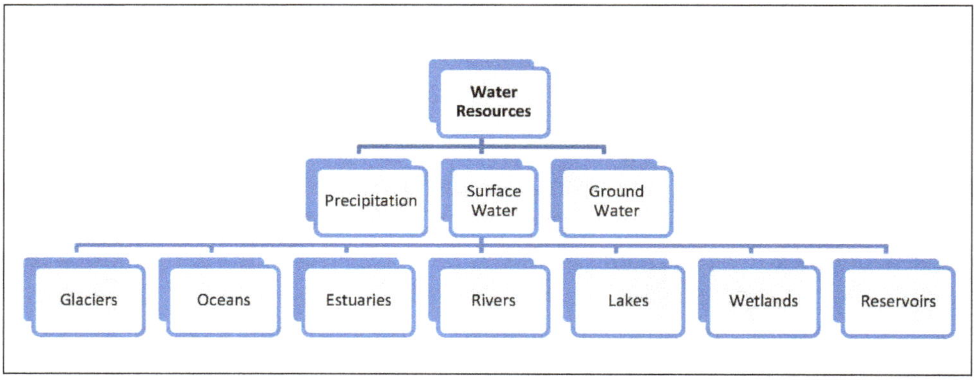

Types of Water Resources

Accessing the water you need might be as simple as turning on the faucet, but this isn't ultimately where your water comes from. In fact, the Earth's water supply comes from a series of different places—some you might expect, and others you might not. To fully understand the process of obtaining quality drinking water for your home, it's crucial that you first know these sources. These are the different types of water sources around the globe and how they each play a role in what comes out of your home's sink.

Surface Water Resources

Surface water resources are the most commonly used method of supplying water to various regions in the United States. This classification primarily includes rivers, lakes, streams, reservoirs, and wetlands—all of which contain fresh water rather than saltwater. These sources are easiest to filter, so they produce the highest-quality drinking water for the general public. Plus, another reason we mostly use these resources is their accessibility—many people live near large lakes or streams from which they can easily extract water. Surface water is therefore the most reasonable option for providing homes and businesses with the resources they need to function. However, people commonly use rivers and lakes for recreational activities such as swimming and fishing; these places also play a part in the industrial manufacturing processes. As such, water from these sources requires extensive sterilization before it's ready for consumption and use.

Groundwater Resources

Believe it or not, there's actually a larger source of water underneath your feet than there is in all the rivers and lakes combined. However, we rarely get to tap into these sources due to how difficult they are to reach. Groundwater fills the cracks in bedrock and sand beneath the surface, making contaminants tedious to filter out in large quantities. These sources also saturate soil and contain so much sediment that the water must undergo a thorough filtration process to even become drinkable. So, while groundwater is the main source of plant hydration, it's not often a sustainable option for people. Fortunately, we aren't completely cut off from groundwater sources—many of them feed some of our surface water supplies through underground springs.

Stormwater Resources

Otherwise known as runoff or rainwater, stormwater is water that comes from heavy weather such as rain, snow, and hail. This water flows over the land and, in the process, collects a variety of pollutants such as engine oil, fertilizer, and pesticides. As it picks up these contaminants, it eventually gathers in different areas, potentially combining with some of our other water sources. For this reason, stormwater—and water from any other type of source—must undergo a series of tests that properly identify and filter out dangerous toxins. In addition, since the majority of this water flows back into the oceans, capturing it beforehand is a great way to increase our overall water supply on land. For this reason, many sustainability experts have researched different ways to collect this water and filter it before it washes away.

Wastewater Resources

You might not initially think of it as an option, but wastewater is another type of water source in the world. This is the water we use for our household, manufacturing, and agricultural activities; it's then disposed of through our drains and local sewage systems. Because this water has already been used, it may contain several potentially toxic elements that must be filtered out and disposed of before the water can be used again. Unfortunately, while recycling water is a common practice in various communities, most wastewater still gets dumped in local surface water resources. This contaminates them and makes it even more difficult to filter out all the contaminants. For this reason, conservation efforts to stop businesses from dumping wastewater into lakes and rivers are on the rise. Preventing this practice better maintains the amount of water for us to live off of.

Saltwater Resources

It's common knowledge that our oceans make up over 70 percent of the planet. However, the salty, abrasive nature of this water makes using it for any of our current processes extremely difficult. In fact, the amount of salt present in ocean water makes it impossible for us to safely drink it in large enough quantities to survive. This is why we dominantly rely on freshwater sources to supply us with the water we need to drink. Fortunately, recent advances in filtration technology have yielded more effective ways to dilute saltwater and remove the acidity that prevents us from using it. Still, desalination plants are low in number due to the amount of energy this filtration process requires. Further evolution of these tools will make the process more sustainable and easier to repeat.

Ice Cap Water Resources

It's theoretically possible for us to retrieve some water from the polar ice caps and glaciers. These large bodies of ice float through the oceans, but they actually consist of fresh water. This makes them some of our most ideal resources—if we can develop reliable ways to tap into them. Unfortunately, the glaciers are too far away for us to regularly utilize, and we have yet to come up with an effective way to melt them. The process of even reaching these territories is too much of an economic burden to be sustainable for long periods. In addition to this, the polar ice caps are crucial to regulating the Earth's surface temperature. Going through the effort of melting them would ultimately throw our global temperatures out of balance and do more harm than good.

Hydrologic Cycle

Water cycle, also called hydrologic cycle is the cycle that involves the continuous circulation of water in the Earth-atmosphere system. Of the many processes involved in the water cycle, the most important are evaporation, transpiration, condensation, precipitation, and runoff. Although the total amount of water within the cycle remains essentially constant, its distribution among the various processes is continually changing.

Evaporation

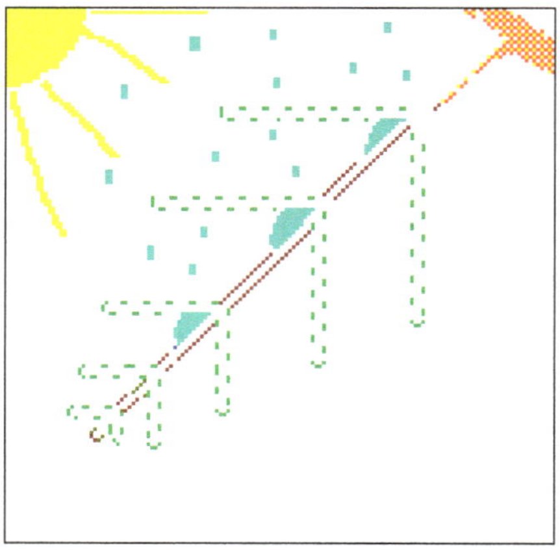

Evaporation occurs when the physical state of water is changed from a liquid state to a gaseous state. A considerable amount of heat, about 600 calories of energy for each gram of water, is exchanged during the change of state. Typically, solar radiation and other factors such as air temperature, vapor pressure, wind, and atmospheric pressure affect the amount of natural evaporation that takes place in any geographic area. Evaporation can occur on raindrops, and on free water surfaces such as seas and lakes.

It can even occur from water settled on vegetation, soil, rocks and snow. There is also evaporation caused by human activities. Heated buildings experience evaporation of water settled on its surfaces. Evaporated moisture is lifted into the atmosphere from the ocean, land surfaces, and water bodies as water vapor. Some vapor always exists in the atmosphere.

Condensation

Condensation is the process by which water vapor changes it's physical state from a vapor, most commonly, to a liquid. Water vapor condenses onto small airborne particles to form dew, fog, or clouds. The most active particles that form clouds are sea salts, atmospheric ions caused by lightning, and combustion products containing sulfurous and nitrous acids. Condensation is brought about by cooling of the air or by increasing the amount of vapor in the air to its saturation point. When water vapor condenses back into a liquid state, the same large amount of heat (600 calories of energy per gram) that was needed to make it a vapor is released to the environment.

Precipitation

Precipitation is the process that occurs when any and all forms of water particles fall from the atmosphere and reach the ground. There are two sub-processes that cause clouds to release precipitation, the coalescence process and the ice-crystal process. As water drops reach a critical size, the drop is exposed to gravity and frictional drag. Falling drop leaves a turbulent wake behind which allows smaller drops to fall faster and to be overtaken to join and combine with the lead drop. The other sub-process that can occur is the ice-crystal formation process. It occurs when ice develops in cold clouds or in cloud formations high in the atmosphere where freezing temperatures occur. When nearby water droplets approach the crystals some droplets evaporate and condense on the crystals. The crystals grow to a critical size and drop as snow or ice pellets. Sometimes, as the pellets fall through lower elevation air, they melt and change into raindrops.

Precipitated water may fall into a water body or it may fall onto land. It is then dispersed several ways. The water can adhere to objects on or near the planet surface or it can be carried over and through the land into stream channels, or it may penetrate into the soil, or it may be intercepted by plants. When rainfall is small and infrequent, a high percentage of precipitation is returned to the atmosphere by evaporation.

The portion of precipitation that appears in surface streams is called runoff. Runoff may consist of component contributions from such sources as surface runoff, subsurface runoff, or ground water runoff. Surface runoff travels over the ground surface and through surface channels to leave a catchment area called a drainage basin or watershed. The portion of the surface runoff that flows over the land surface towards the stream channels is called overland flow. The total runoff confined in the stream channels is called the stream-flow.

Interception

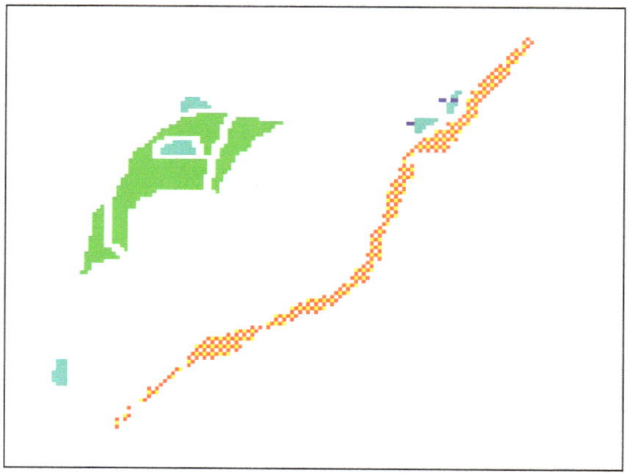

Interception is the process of interrupting the movement of water in the chain of transportation events leading to streams. The interception can take place by vegetal cover or depression storage in puddles and in land formations such as rills and furrows. When rain first begins, the water striking leaves and other organic materials spreads over the surfaces in a thin layer or it collects at points or edges. When the maximum surface storage capability on the surface of the material is exceeded, the material stores additional water in growing drops along its edges. Eventually the weight of the drops exceeds the surface tension and water falls to the ground. Wind and the impact of rain drops can also release the water from the organic material. The water layer on organic surfaces and the drops of water along the edges are also freely exposed to evaporation.

Additionally, interception of water on the ground surface during freezing and sub-freezing conditions can be substantial. The interception of falling snow and ice on vegetation also occurs. The highest level of interception occurs when it snows on conifer forests and hardwood forests that have not yet lost their leaves.

Infiltration

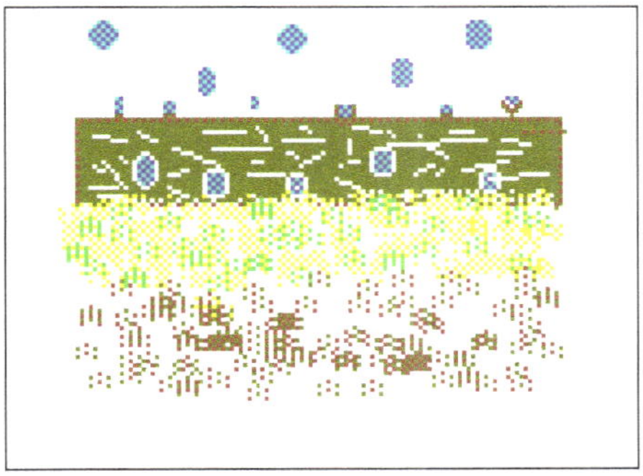

Infiltration is the physical process involving movement of water through the boundary area where the atmosphere interfaces with the soil. The surface phenomenon is governed by soil surface conditions. Water transfer is related to the porosity of the soil and the permeability of the soil profile. Typically, the infiltration rate depends on the puddling of the water at the soil surface by the impact of raindrops, the texture and structure of the soil, the initial soil moisture content, the decreasing water concentration as the water moves deeper into the soil filling of the pores in the soil matrices, changes in the soil composition, and to the swelling of the wetted soils that in turn close cracks in the soil. Water that is infiltrated and stored in the soil can also become the water that later is evapotranspiration or becomes subsurface runoff.

Percolation

Percolation is the movement of water though the soil, and its layers, by gravity and capillary forces. The prime moving force of groundwater is gravity. Water that is in the zone of aeration where air exists is called vadose water. Water that is in the zone of saturation is called groundwater. For all practical purposes, all groundwater originates as surface water. Once underground, the water is moved by gravity. The boundary that separates the vadose and the saturation zones is called the water table. Usually the direction of water movement is changed from downward and a horizontal component to the movement is added that is based on the geologic boundary conditions.

Geologic formations in the earth's crust serve as natural subterranean reservoirs for storing water. Others can also serve as conduits for the movement of water. Essentially, all groundwater is in motion. Some of it, however, moves extremely slowly. A geologic formation which transmits water from one location to another in sufficient quantity for economic development is called an aquifer. The movement of water is possible because of the voids or pores in the geologic formations. Some formations conduct water back to the ground surface. A spring is a place where the water table reaches the ground surface. Stream channels can be in contact with an unconfined aquifer that approaches the ground surface. Water may move from the ground into the stream, or visa versa, depending on the relative water level. Groundwater discharges into a stream forms the base flow of the stream during dry periods, especially during droughts. An influent stream supplies water to an aquifer while and effluent stream receives water from the aquifer.

Transpiration

Transpiration is the biological process that occurs mostly in the day. Water inside of plants is transferred from the plant to the atmosphere as water vapor through numerous individual leave openings. Plants transpire to move nutrients to the upper portion of the plants and to cool the leaves exposed to the sun. Leaves undergoing rapid transpiration can be significantly cooler than the surrounding air. Transpiration is greatly affected by the species of plants that are in the soil and it is strongly affected by the amount of light to which the plants are exposed. Water can be transpired freely by plants until a water deficit develops in the plant and it water-releasing cells (stomata) begin to close. Transpiration then continues at a must slower rate. Only a small portion of the water that plants absorb are retained in the plants.

Vegetation generally retards evaporation from the soil. Vegetation that is shading the soil, reduces the wind velocity. Also, releasing water vapor to the atmosphere reduces the amount of direct evaporation from the soil or from snow or ice cover. The absorption of water into plant roots, along with interception that occurs on plant surfaces offsets the general effects that vegetation has in retarding evaporation from the soil. The forest vegetation tends to have more moisture than the soil beneath the trees.

Runoff

Runoff is flow from a drainage basin or watershed that appears in surface streams. It generally consists of the flow that is unaffected by artificial diversions, storages or other works that society might have on or in a stream channel. The flow is made up partly of precipitation that falls directly on the stream, surface runoff that flows over the land surface and through channels, subsurface runoff that infiltrates the surface soils and moves laterally towards the stream, and groundwater runoff from deep percolation through the soil horizons. Part of the subsurface flow enters the stream quickly, while the remaining portion may take a longer period before joining the water in the stream. When each of the component flows enters the stream, they form the total runoff. The total runoff in the stream channels is called stream-flow and it is generally regarded as direct runoff or base flow.

Storage

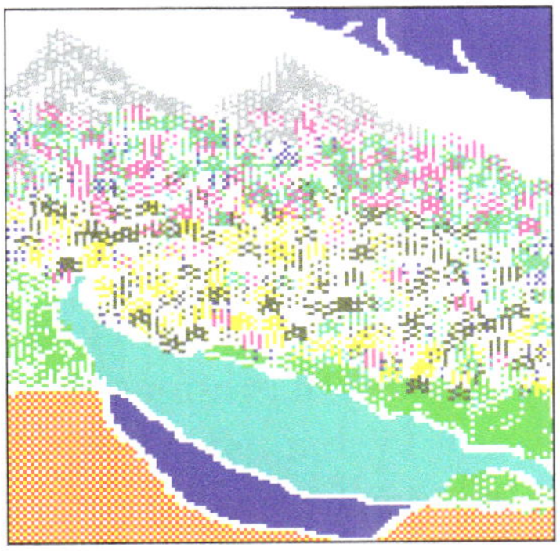

There are three basic locations of water storage that occur in the planetary water cycle. Water is stored in the atmosphere; water is stored on the surface of the earth, and water stored in the ground. Water stored in the atmosphere can be moved relatively quickly from one part of the planet to another part of the planet. The type of storage that occurs on the land surface and under the ground largely depend on the geologic features related to the types of soil and the types of rocks present at the storage locations. Storage occurs as surface storage in oceans, lakes, reservoirs, and glaciers; underground storage occurs in the soil, in aquifers, and in the crevices of rock formations.

The movement of water through the eight other major physical processes of the water cycle can be erratic. On average, water the atmosphere is renewed every 16 days. Soil moisture is replaced about every year. Globally, waters in wetlands are replaced about every 5 years while the residence time of lake water is about 17 years. In areas of low development by society, groundwater renewal can exceed 1,400 years. The uneven distribution and movement of water over time, and the spatial distribution of water in both geographic and geologic areas, can cause extreme phenomena such as floods and droughts to occur.

Groundwater Hydrology

Ground-water hydrology is a branch of hydrology that deals with the existence, movement, and quality of undergroundwater beneath the Earth's surface. As per the U.S. National Research Council, hydrology is defined as, "The science that treats the water of the Earth, and their occurrence, circulation, and distribution, their physical and chemical properties, and their reaction with the environment, including their relation to living things". Life originated in water and depends on water. The entire living world of our planet, i.e., plants, animals and human beings depend upon water for their survival. Life is unthinkable without water. Most of the societies and cultures originated near water sources. Water is required for each and every step of life for fulfilment of their various purposes like drinking, irrigation, transportation, and recreation. Water present beneath the land surface is counted as subsurface water or undergroundwater, which is usually differentiated into two different zones i.e. saturated zone and unsaturated zone. In most of the areas, unsaturated zone exists immediately below the land surface and usually found above the saturated zone. This zone of underground water contains both air and water. However, saturated zone contains interconnected openings which are saturated in water. This zone of underneath land surface actually presents true groundwater, whose existence is in pore spaces and fractures of sediments beneath the Earth's surface. Therefore, in other sense, true groundwater is found in the open space between sedimentary materials or in fractured rocks, existed beneath the unstaurated zone, and the available water has a pressure greater than atmospheric pressure. Moreover, groundwater facilitates availability of water for human consumption. Groundwater is imitative from several sources, however, primarily rainfall and snow melting acts as precursor for the saturated zone of groundwater. The hydrological cycle is the central focus for groundwater hydrology. A part of water percolates from the land surface and infiltrate to saturated zone of groundwater through unsaturated zone. Therefore, unsaturated zone has a great importance to groundwater hydrology and usually differentiated into three different sub-zones: soil zone, intermediate zone, and capillary fringe. The capillary fringe represents the zone exists immediately above the water table, where some or all of the interstices are filled with water at pressure lower than atmospheric pressure. This sub-zone of unsaturated zone remains continuous with water below the water table.

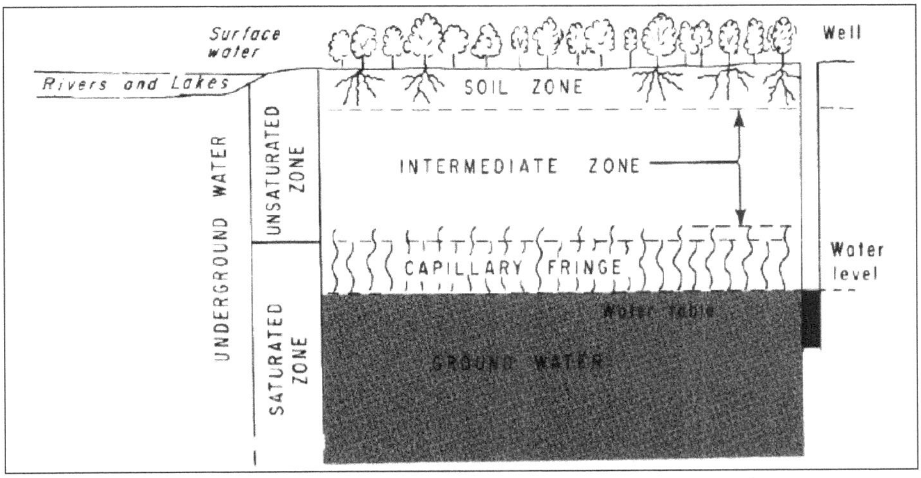

Zonation of underground water underneath the Earth surface.

Some facts about groundwater:

- Water is most widely occurring substance on Earth; however, only 2.8% of total water resources exist as fresh water. Moreover, the biggest portion of fresh water remains locked in glaciers and icebergs or lies in deep underground aquifers. Out of this 2.8%, about 0.6% is available as groundwater and only 0.3% can be extracted economically with the present drilling technologies.
- The unseen ocean of fresh water beneath the Earth surface makes about 35 times the available water in lakes and streams at any time.
- The available groundwater for human consumption generally occurs within 800 meters from the Earth surface.
- Currently, around one fifth of total water consumed at world level, is maintained by groundwater resources only and 80% of the total groundwater consumption is meant for irrigating agricultural lands. Moreover, entire Earth surface can be covered with a 55 meter thick layer from the volume of groundwater available beneath the Earth surface.
- The groundwater moves with extremely slow speed underneath the Earth surface through the inter-particle space or between the networks of fractures of

unconsolidated material. The rate of exchange of groundwater in 280 years is almost 1/9,000 the rate of exchange of fresh water in rivers and streams.

- The movement of groundwater generally follows the path from recharge areas, from where water percolates to the ground, to the discharge areas, from where it exits from the ground.

Table: The occurrence of water resources around the globe.

Water resources	Volume (1,000 Km³)	Percentage
Atmospheric water	13	0.001
Surface water		
Oceans saline water	1,320,000	97.2
Sea saline water	104	0.008
Lake fresh water	125	0.009
Rivers fresh water	1.25	0.0001
Biosphere fresh water	29,000	2.15
Glacial water	50	0.004
Groundwater		
Water in Vadose zone	67	0.005
Groundwater in the depth (less than 800 meters)	4,200	0.31
Groundwater in the depth (more than 800 meters)	4,200	0.31
Total	13,60,000	100

Classification of Water with Reference to its Origin

The relationship between different generic types of water is depicted in figure. Most of water resources are segregated into different generic types on the basis of their origin and occurrence which are described as under:

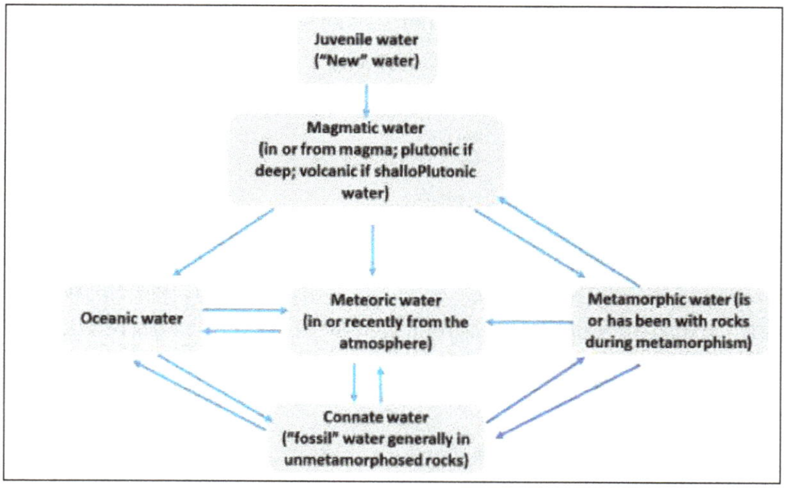

Relationships between different genetic types of water on the basis of their occurrence.

Magmatic Water

The occurrence of water within, and in equilibrium with, magma or water-rich volatile fluids derived from magma, contributes to magmatic water. The term Plutonic water is applied, where the separation is deep, while water from shallow depths called volcanic water. As far as water availability is concerned, it is of only theoretical importance and can be released to the atmosphere during a volcanic eruption.

Juvenile Water

This form of water is also known as new water, which has not been introduced to hydrosphere previously. The magmatic water is also a part of juvenile water, which may be either contributed from deep seated magma or shallow volcanic eruption.

Meteoric Water

Most of the meteoric water is as a resultant of different forms of atmospheric precipitation, usually a part of present day hydrological cycle. This resource facilitates the water available in lakes, rivers, streams, wells and ice melts, which originates directly or indirectly from the precipitation. Although a large volume of precipitated water or water from melting of snow and ice, reaches the seas and oceans through the surface flows or runoffs, however, a considerable portion is also assessable to groundwater. The water in filtered remain continue its downward journey till it reaches to zone of saturation.

Connate Water

The water retained in ancient aquifers, encountered at great depth in sedimentary rocks, either through marine or fresh water in origin, contributes to connate water. Generally, this type of water is present in hydraulic discontinuity with hydrological cycle and is also coined with a misnomer term, fossil water. Connate water is usually linked with fossil fuels including oil or gases, which are marine in origin. Sometimes, incomplete compaction may cause retention of small quantity of water with rocks like limestone, sandstone, and gravel.

Metamorphic Water or Rejuvenated Water

This term is used for water associated with hydrous rocks like micas, clay, etc., during the process of metamorphism. This genetic type of groundwater encounters more of academic interest rather than a water resource.

Hydrological Budget

The global hydrological cycle maintains the continuous churning of water on, above and below the Earth surface, supporting several life forms on this planet. The system can be subdivided usually into three different subsystems, including atmospheric water system, surface water system and subsurface water system. The atmospheric system includes precipitation, evaporation, interception, and transpiration, while surface water system includes overhead flow, surface run-off and run-off to streams, seas and oceans. The subsurface water system includes infiltration, subsurface flow, and groundwater recharge and groundwater flow. Under normal conditions, the travel time of groundwater can varies from less than a day to more than a million years. The hydrological budget is very important and includes inflows, outflows, and storage of water as shown in the following equations.

Surface water system hydrological budget: $P + Q_{in} - Q_{out} + Q_g - E_s - T_s - I = \Delta S_s$

Surface groundwater system hydrological budget: $I + G_{in} - G_{out} - Q_g - E_g - T_g = \Delta S_g$

System hydrological budget:

$$P - (Q_{out} - Q_{in}) - (E_s - E_g) - (T_s - T_g) - (G_{out} - G_{in}) = \Delta(S_s - S_g)$$

Where:

- P = Precipitation
- Q_{in} = Surface water flow into the system
- Q_{out} = Surface water flow out of the system
- Q_g = Groundwater flow into the stream
- E_s = Surface evaporation
- T_s = Transpiration
- I = Infiltration
- ΔS_s = Change in water storage of the surface water system
- G_{in} = Groundwater flow into the system
- G_{out} = Groundwater flow out of the system
- ΔSg = Change in groundwater storage
- E_g = Evaporation in groundwater system
- T_g = Transpiration in groundwater system

Rock Properties Affecting Groundwater

Different hydraulic properties of rocks and soil, including porosity, soil classification, and specific surface, have a considerable role towards controlling occurrence, availability and flow of groundwater.

Soil Classification

The amount of groundwater recharge, storage, discharge, and the extent of groundwater contamination depends on the soil properties (like death, texture, porosity, specific yield, permeability, filtering capacity). The soil serves the purpose of attenuation on filtering solid particles, chemicals and dissolved substances. Microbial and biological processes transform such contaminants and retard such substances. The coarse-textured material (like sand), are more permeable and appears the worst for contaminant removal, whereas fine-textured soils are the best to meet the purpose due to less permeability and higher contact between soil and contaminants.

Specific Surface and Specific Yield

The surface area depends on particle size, shape and mineral types and generally refers to the area per unit weight of the material (m^2/g).In saturated zone of groundwater, water may adhere to the particles due to cohesion, or adhesion and surface tension, forming a film around particles. Such groundwater in saturated rocks or soil gives an idea about specific yield, which is the ratio of volume of water that could drain from saturated rock to the total volume of rock. The value of specific yield is always less than porosity. However, likewise porosity, specific yield is also influenced with grain size.

Porosity

The soil or rock fractions occupied spaces between the grains, which serve as space (voids, interstices, pores, or pore space) for the existence of groundwater. These opening are of great interest for hydrologists, as these space serve as water conduits. The opening of rocks acts as a reservoir for groundwater. In unconsolidated rocks (like gravel and sand), opening occur between the particles of rocks or mineral grains, which are known with the term primary openings. Whereas, the purpose is facilitated with fractures in consolidated rocks (like basalt sandstone, granite, etc.) and is termed as secondary openings. The volume of opening in the rock or underneath soil determines the quantum of groundwater. The ratio of total volume of openings/spaces to the total volume of rocks or soil, represent the porosity.

Permeability

The rock particles, fracture size and shape play an important role in determining the porosity and permeability, which in turn regulate the aquifer characteristics and hydraulic conductivity of groundwater. The permeability is the parameter which determines the ability of water transmission and movement, and facilitates availability of water for human consumption. Coarse materials are generally more permeable; however, large openings may lower down the permeability because of filling of opening with fine grains. On other hand, granular soil (like clay), offers more fraction between the grains, which in turn reduce its permeability. In rock with fractures, permeability can be determined with opening size, interconnectivity between fractures and porosity. Usually, crystalline rocks are less permeable due to fewer openings and less porosity. However, a volcanic rock with broader openings enables faster movement of water.

Aquifers

Aquifers refer to any underground layer of water-bearing rock or geological formation that yields sufficiently groundwater for wells and springs. According to geological terms an Aquifer could be referred to as a body of saturated rock or geological formation through which water can easily move (permeability) into wells and streams. The top of the water level in an aquifer is called the water table. An aquifer fills with water from rain or melted snow that drains into the ground. In certain areas, water could pass through the soil of the aquifer while in other areas it enters through joints and cracks in rocks where it moves downwards until it encounters rocks that are less permeable. Aquifers generally are known to serve as reservoirs and could dry up when people drain them faster than they are been refilled by nature.

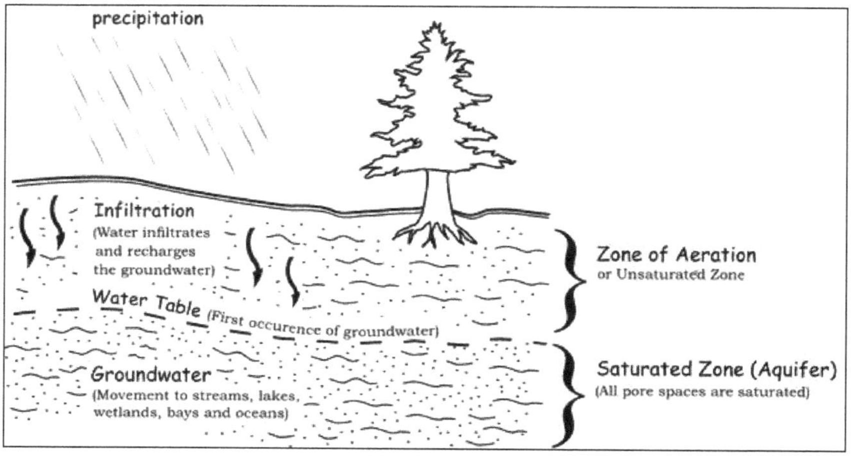

Aquifer formation.

Aquifers must not only be permeable but must also be porous and are found to include rock types such as sandstones, conglomerates, fractured limestone and unconsolidated sand, gravels and fractured volcanic rocks (columnar basalts). While some aquifers have high porosity and low permeability others have high porosity and high productivity. Those with high porosity and low permeability are referred to as poor aquifers and include rocks or geological formation such as granites and schist while those with high porosity and high permeability are regarded as excellent aquifers and include rocks like fractured volcanic rocks.

Classification of Aquifers

Aquifers are generally been classed into two main categories namely confined aquifer and unconfined aquifers.

Confined Aquifers

Confined Aquifers are those bodies of water found accumulating in a permeable rock and are been enclosed by two impermeable rock layers or rock bodies. Confined Aquifers are aquifers that are found to be overlain by a confining rock layer or rock bodies, often made up of clay which might offer some form of protection from surface contamination. The geological barriers which are non-permeable and found exist between the aquifer causes the water within it to be under pressure which is comparatively more than the atmospheric pressure. The presence of fractures, or cracks in bedrocks is also capable of bearing water in large openings within bedrocks dissolving some of the rock and accounts for high yields of well in karst terrain counties like Augusta, Bath within Virginia. Groundwater flow through aquifers is either vertically or horizontally at rates often influenced by gravity and geological formations in these areas.

Confined aquifers could also be referred to as "Artesian aquifers" which could be found most above the base of confined rock layers. Punctured wells deriving their sources from artesian aquifers have fluctuation in their water levels due more to pressure change than quantity of stored water. The punctured well serve more as conduits for water transmission from replenishing areas to natural or artificial final points. In terms of storativity, confined aquifers have very low storativity values of 0.01 to 0.0001.

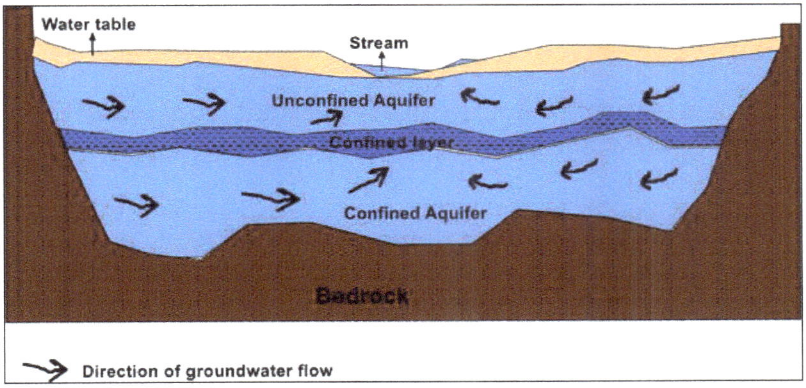

Schematic cross-section of aquifer types.

Unconfined Aquifers

Unconfined Aquifers unlike confined aquifers are generally found located near the land surface and have no layers of clay (or other impermeable geologic material) above the water table although they are found lying relatively above impermeable clay rock layers. The uppermost boundary of groundwater within the unconfined aquifer is the water table, the groundwater in an unconfined aquifer is more vulnerable to contamination from surface pollution as compared to that in confined aquifers this been so due to easy groundwater infiltration by land pollutants. Fluctuation in the level of groundwater varies and depends on the stored up groundwater in the space of the aquifer which in turn affects the rise or fall of water levels in wells that derive their source from aquifers. Unconfined aquifers have a storative value greater than 0.01. “Perched aquifers” are special cases of unconfined aquifers occurring in situation where groundwater bodies are separated from their main groundwater source by relatively impermeable rock layers of small areal extents and zones of aeration above the main body of groundwater The quantity of water found available in this type of aquifer is usually minute and available for short periods of time.

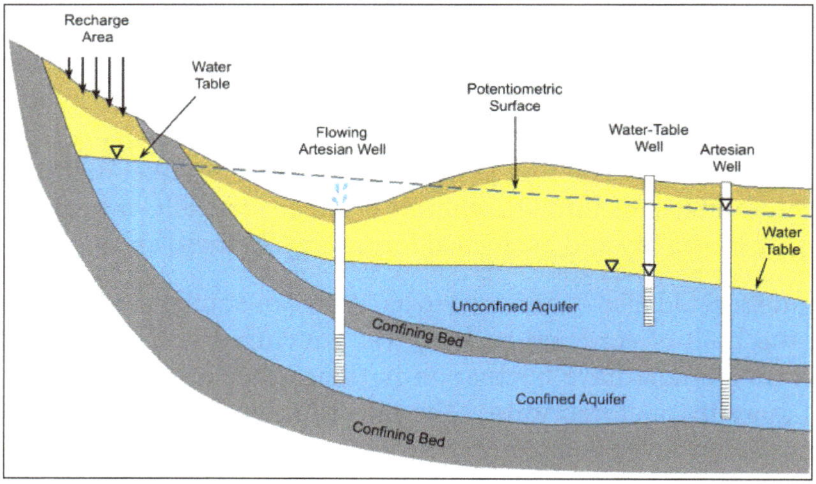

Schematic cross-section of aquifer types.

Petro-Physical Properties of Aquifers

Petro-physical properties of aquifers are properties that help in the defining and characterizing aquifers. Some of the properties considered are:

Hydraulic Conductivity

Hydraulic Conductivity could be described as the relative ease with which a fluids (groundwater) flows through a medium (in this case a geological formation or rock) which is quite different from intrinsic permeability in that though it describes the water-transmitting property of the medium it is however not influenced by the temperature, pressure or the fluid passing through the geological formation. Hydraulic conductivity of a soil or rock or geological formation depends on a variety of physical factors amongst which includes porosity, particle size and distribution, arrangement of particles and other factors.

Mathematically hydraulic conductivity could be defined by the formula below:

$$K=k\frac{\rho g}{\mu}$$

Where, K is the hydraulic conductivity (cm/s or m/s), k is the intrinsic permeability, ρ is the density of fluid, μ is the dynamic viscosity of fluid.

Seconds (s) could be converted to days by time conversions.

Generally, for unconsolidated porous media, hydraulic conductivity varies with particle size as such clayey materials exhibits low values of hydraulic conductivity as compared to sands and gravels that exhibits high values of hydraulic conductivity (150 m/day for coarse gravels, 45 m/day for coarse sand and 0.08 m/day for clay). This is so because the small particle size arrangements (fine grained) in geological formations contained mainly of clayey materials though porous is not permeable enough to allow groundwater flow within it however in sands and gravels (medium to coarse grained) we have medium to coarse arrangement of particle sizes which results to a porous and permeable geological formations or rocks that allows a higher ease of groundwater flow. It is however essential to point out that we could have geological formations or rock that exhibit medium values of hydraulic conductivity, this is in the case where you have a geological formation made up of moderate amounts of clayey material and sandy materials. It should also be noted that variations in hydraulic conductivity values of geological formations or rocks is dependent on factors such as weathering, fracturing, solution channels and depth of burial.

Porosity

Porosity of a geological formation or rock or soil could be described as the measure of the contained voids or interstices expressed as a ratio of the volume of voids to the total volume. It could also be defined as the volume of pores within a rock or soil sample divided by the total volume of the rock matrix (pores and solid materials contained with the rock). When a rock is emplaced by either cooling from an igneous melt or induration from loose sediment or soil formation from weathering of rock materials, it possess an inherent porosity known as primary porosity which reduces with time by actions of cementation or compaction. However, when joints, fissures, fractures or solution cavities formed within rocks after the must have been emplaced it is referred to as secondary porosity. Therefore, total porosity is the sum of primary and secondary porosities.

If all the pores found contained in a rock are not connected, then only a certain fraction of the pores would allow for water movement. The fraction that allows for water movement is known as the effective porosity example of which includes pumice, glassy volcanic rock (solidified froth) probably would float in water because its total porosity is high and it contains much entrained gas.

Porosity of a rock is determined to a large extent by the packing arrangement of particle sizes and the uniformity of its grain-size distribution. As such a cubic packing would give a porosity of 47.65%, the greatest and most ideal a rock with uniform spherical grains can achieve as the centers of eight such grains from vertices of a cube. However, if the packing arrangement of the rock where to change to be that of a rhombohedral then, its porosity would reduce to 25.85% as the centers of the eight adjacent spheres form the vertices of rhombus.

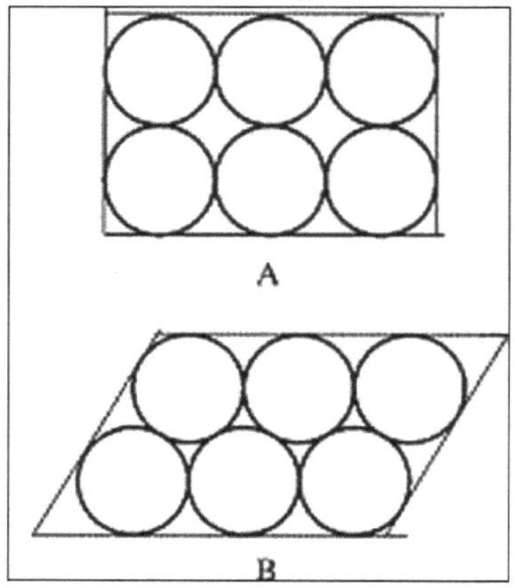

(A) Cubic packing (B) Rhombohedral packing.

Mathematically porosity (n) is given by the formula below:

$$\text{Porosity}(n) = \frac{V_p}{V_t}$$

Where, V_p is the pores of rock or soil sample, V_t is the total volume of pores and solid material.

Transmissivity

Transmissivity (T) more simply could be defined as the property of aquifer to transmit water. It could also be defined as the amount of water that can be transmitted horizontally through an aquifer unit by full saturated thickness of the aquifer under a hydraulic gradient of 1 or as the rate at which water of prevailing kinematic viscosity is transmitted through a unit width of aquifer under a unit gradient.

It could be mathematically defined as:

$$T = Kb\left(m^2/\text{day}\right)$$

Where, T is the transmissivity, K is the hydraulic conductivity, b is the saturated thickness of the aquifer.

Aquifers are characterized by petro-physical properties such as hydraulic conductivity (alternatively called permeability), transmissivity (product of hydraulic conductivity and aquifer thickness) and diffusivity (ratio of transmissivity and storage coefficient). These properties could be examined using geophysical techniques such as Electrical Resistivity, Seismic techniques.

Aquifer Characterization – Electrical Techniques

In geophysical investigations using electrical techniques, two primary properties of interest been considered are electrical conductivity or dielectric constant.

Electrical techniques make involves profiling and sounding mode of data collection which both involves the use of direct currents or low frequency altering currents passing through the subsurface. While profiling use of the terms profiling involves the inferring of subsurface measurements based on lateral changes in electrical properties over constant subsurface, Sounding infers subsurface measurements of single location as a function of changes in petro-physical properties as function of depth. The use of either of the two modes for data collection is function of the purpose of the investigation.

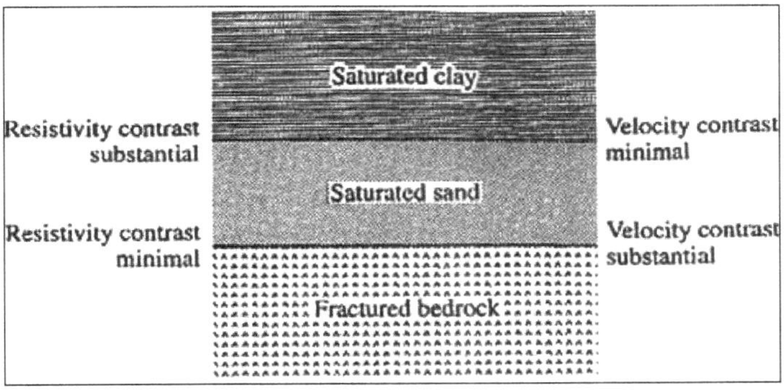

Schematic geological section and associated resistivity and velocity contrasts at interfaces.

Electrical Resistivity

Electrical resistivity (ER) is more frequently been used as compared to other electrical techniques in groundwater investigations of which includes the characterization of aquifers. Electrical Resistivity (ER) involves the introduction of time – varying direct current (DC) or very low frequency (<1 Hz) current into the ground between two current electrodes to generate potential differences as measured at the surface with units of Ohm-meters (Ω-m). A deviation from the norm in the pattern of potential differences expected from homogeneous proffer the necessary information on the form and electrical properties of subsurface in homogeneities.

A typical Electrical Resistivity (ER) investigation made up of a 2 – electrode system would include 2 – current electrodes and 2 – potential electrodes. As current is been injected into the ground, corresponding potential differences (ΔV) is measured. This measurement coupled with known current (I) and a geometric factor (K) that is a function of the particular electrode configuration, can be used to calculate resistivity (ρ) following Ohm's law:

$$\rho=\frac{\Delta V}{I}K$$

The expression in previous equation for a homogeneous ground is also the same applied for heterogeneous ground; however the general term "apparent resistivity (ρa)" is substituted for resistivity (ρ) in previous equation. Apparent resistivity (ρa) is used here rather than the actual resistivity of the subsurface due to the non-homogeneity nature of the subsurface.

A four – electrode configurations is been used most commonly when it comes to measuring apparent resistivity of the subsurface. The simplest of these configurations is the Wenner configuration where the outer two current electrodes C_1 and C_2, apply a constant current, and the inner two potential electrodes, labeled P_1 and P_2, measures voltage difference created by this current. The electrode spacing has a fixed value a, and the apparent resistivity of the subsurface sampled by this array could be computed using the equation:

$$\rho a = \frac{\Delta V}{I} 2\pi a$$

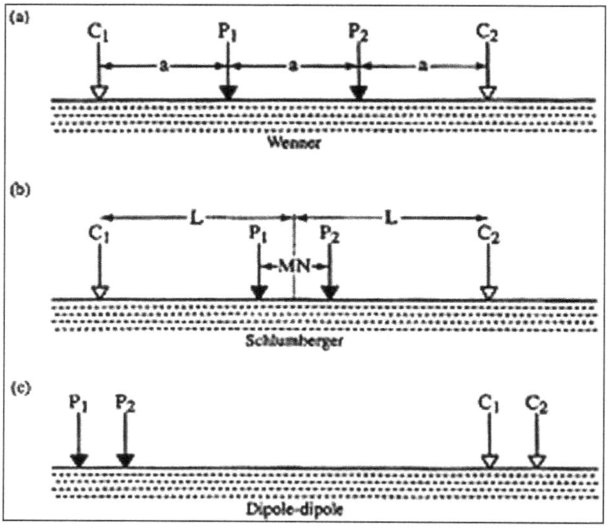

Common electrode configuration used to measure apparent resistivity of the subsurface C_1 and C_2 are the current electrodes and P_1 and P_2 are the potential electrodes. (a) Wenner Array (b) Schlumberger array (c) Dipole – dipole array.

Asides the Wenner array mode of electrode configuration, another commonly used electrode configuration is the Schlumberger array, where the spacing (MN) between the potential electrodes (P_1, P_2) is much smaller as compared to the spacing (2 L) between the current electrodes (C_1, C_2).

The electrode configuration in represents that of a dipole – dipole array where the potential electrode pair and current electrode is closely spaced, however there exist significant distances between the two sets of electrodes. Unlike the cases of the Wenner and Schlumberger arrays, where data collected through either profiling or sounding mode depends a lot on the electrode array geometry.

To illustrate, we consider the Wenner array. Profiling involves the lateral movement of the entire array along the surface at fixed distances to obtain apparent resistivity measurements as a function of distance. The values of the measurements are assigned to the geometric center of the electrode array. Interpretation of measurements is usually with its data aimed at location of geological structures buried stream channels, aquifers or water bearing formations etc.

Sounding unlike profiling involves gradual and progressively expansion of expansion of the array about a fixed central point with current and potential electrodes being maintained at a relative spacing with depth been a function of electrode spacing and subsurface resistivity contrasts. The dashed lines representing current flow lines in an homogeneous environment while bold lines represents actual current flow in single interface that separates units with different resistivities. Next we look how electrode spacing, current and its' influence on depth of penetration. In figure, when the electrode spacing is close, it is observed that the current only upper interface (i.e. the interface of lower resistivity). The scenario in figure is different; as electrode spacing has increased resulting in greater penetration depth and higher apparent resistivity values due to the influence of the lower (higher resistivity) layer. Lastly when the electrodes are farther apart, only substantial current amounts are found to flow through the resistivity layer.

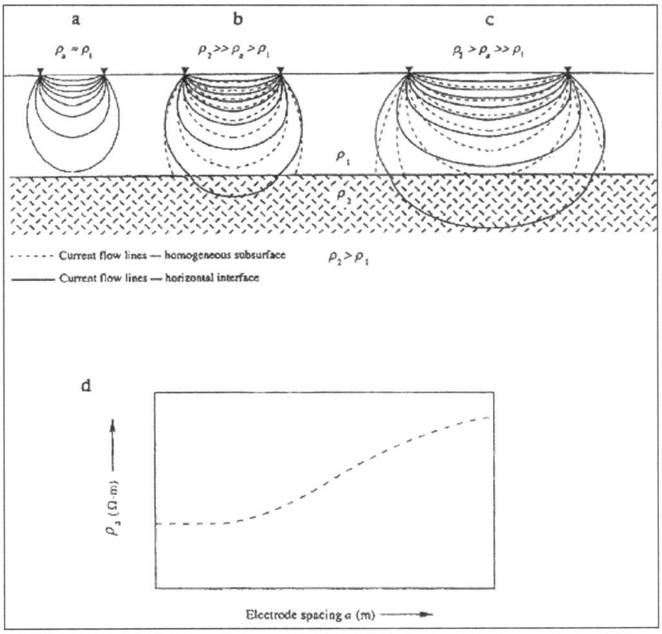

Figure shows effects of electrode spacing and presence of an interface on apparent resistivity measurements. The dashed lines represent current flow lines in the absence of the interface and the solid lines represent actual current flow lines (a–c) as the current electrode spacing is increased, the current lines penetrate deeper and the apparent resistivity measurements are influenced by the lower (more resistive) layer. (d) the qualitative variations in apparent resistivity as a function of electrode spacing are illustrated by the two-layer sounding curve.

The curve reveals qualitative variations in apparent resistivities which increases with electrode spacing, a, this curve is known as a sounding curve revealing geology of the subsurface with resistivity increasing with depth provided geology is homogeneous. However in the case where the geology is inhomogeneous it results in a complex sounding curve whose interpretation is non-unique. To interpret electrical resistivity sounding data, various curve-fitting or computer inversion schemes are used or measured and compared with model computations. A classic example where both modes of data acquisition (profiling and sounding) is been used is in the location of a buried stream channel using Wenner array. The contour map produced from resistivity measurements of several profiles collected near San Jose, CA, using an a-spacing of 6.1 reveals contours of equal apparent resistivity delineating an approximately east–west trending high apparent resistivity values. To understand the cause of high apparent resistivity values here, a geological cross-section (BA) was drawn across the map. The geological cross-section (BA) drawn is based on four expanding – spread traverses (soundings), apparent resistivity profile information and information from three boreholes whose locations are indicated on the cross-section. The critical observation of the cross-section shows that the area with high-resistivity as on the apparent resistivity map is a zone of gravels and boulders that defines the location of a buried stream channel (subsurface structure).

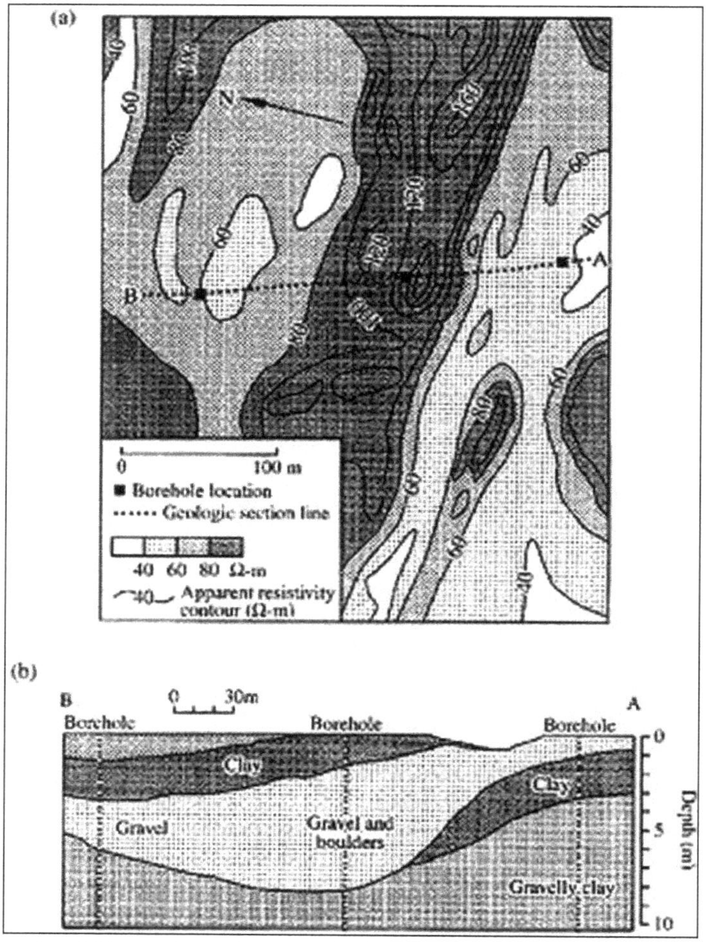

Resistivity survey used to delineate lateral and vertical variations in subsurface stratigraphy. (a) Contour map produced from resistivity measurements, (b) a geologic cross-section (BA) revealing high-resistivity trend in a zone of gravel and boulders that define the location of a buried stream channel.

Aside the mapping of subsurface structure and stratigraphy, electrical resistivity measurements could be channeled towards the inferring lithological information and hydrogeological parameters needed for the mapping groundwater. For groundwater mapping, electrical conduction (inverse of electrical resistivity) is considered. Here the interest is the delineation of connected pore spaces, void spaces, interstices, fractures within rocks that are water filled which leads to a reduced resistivity values and high conductivity. However more information is still needed as high conductivity within rock formation or units could be due to a number of things asides water some of which includes presence of clay minerals, contamination plumes etc.

Common earth materials have wide range of electrical resistivity values revealed in table, however some of these values are known to overlap for different earth materials. Values commonly vary over 12 orders of magnitude and have a maximum range of 24 orders of magnitude. The following statements as regarding electrical resistivity hold:

- Resistivity is sensitive to moisture content; thus unsaturated sediments usually have higher resistivity values than saturated sediments.
- Sandy materials generally have higher resistivity values than clayey materials
- Granitic bedrock generally has a higher resistivity value than saturated sediments and frequently offers a large apparent resistivity contrast when overlain by these sediments.

Table: Resistivity and dielectric constants for typical near-surface materials.

Material	Resistivity (ohm-m)	Dielectric Constant
Sand (Dry)	10^3-10^7	3-6
Sand (Saturated)	10^2-10^4	20-30
Silts	10^2-10^3	5-30
Shales	10-10^3	5-15
Clays	1-10^3	5-40
Humid Soil	50-100	30
Cultivated Soil	200	15
Rocky Soil	1000	7
Sandy Soil (Dry)	7100	3
Sandy Soil (Saturated)	150	25
Loamy Soil (Dry)	9100	3
Loamy Soil (Saturated)	500	19
Clayey Soil (Dry)	3700	2
Clayey Soil (Saturated)	20	15
Sandstone (Saturated)	25	6
Limestone (Dry)	10^9	7
Limestone (Saturated)	40	4-8

Besalt (Saturated)	100	8
Granite	10^3-10^5	4-6
Coal	10^4	4-5
Fresh Water	30-10^4	81
Permafrost	10^2-10^5	4-8
Dry Snow	10^5-10^6	1
Ice	10^3-10^5	4-12

Asides the use of sounding curves, empirical formulae have also been adapted in relating measurement of apparent resistivity with hydrological parameters of interest as this relates to aquifers. The empirical formula developed in the laboratory by Archie relates these parameters:

$$\rho_r = a\, \varnothing^{-m}\, S^{-n}\, \rho_w$$

Where, ρ_r is the electrical resistivity of the rock, ρ_w is the pore water resistivity, $\varnothing$ is the fractional porosity, S is the fractional water saturation.

And n, a, and m are constants {$n \approx 2$, $0.6 \le a \le 1.0$, and $1.4 \le m \le 2.2$; Ward}. Though Archie's law was formulated using lithified materials. The equation presented by previous equation is used generally for well log interpretation however if ρ_r, ρ_w, and $\varnothing$ can be measured separately such that a and m are estimated reasonably then the fractional water saturation could also be inferred using electrical surveys. This concept was utilized by Pfeifer and Anderson to observe and monitor the migration of tracer-spiked water through the subsurface using resistivity array.

It could be said that the complexities that exist in the interpretation of sounding curves and the non-unique solution it gives, suggests the suitability of surface resistivity in determined subsurface geology. Also due to its sensitivity to parameters like moisture content it's been termed a useful tool in hydrological investigations.

Electromagnetic Induction

Electromagnetic (EM) techniques as tool for geophysical exploration has dramatically increased in recent years served as a useful tool for groundwater and environmental site assessment. It involves the propagation of continuous-wave or transient electromagnetic fields in and over the earth through resulting in the generation of time-varying magnetic field. For any of such surveys to be carried out three components are essential; a transmitters, receivers, buried conductors or conductive subsurface. These three form a trio of electric-circuit coupled by an EM induction with currents been introduced into the ground directly or through inductive means by the transmitters.

The Primary field travels from the transmitter coil to the receiver coil via paths above and below the surface. Where a homogenous subsurface is detected no difference is observed between the fields propagated above, below and within the surface other than a slight reduction in amplitude. However, the interaction of the time-varying field with a conductive subsurface induces eddy currents, which gives rise to a secondary magnetic field. The attributes of the fields generated, such as amplitude, orientation and phase shift can be measured by the receiver coil and compared with those of the primary field as such information about the presence of subsurface conductors, or subsurface electrical conductivity distribution can be inferred.

It's paramount to recall that electrical conductivity is an inverse of electrical resistivity; as such electrical conductivity measurements made using electromagnetic methods is also dependent on subsurface texture, porosity, presence of clay minerals, moisture content and the electrical resistivity of the pore fluid presence. The acquisition of EM data requires less time, achieving greater depth of investigation than resistivity techniques. However, the equipment used are expensive and the methods used to qualitatively interpret data from EM surveys is complicated than those used in resistivity methods. This is because a conductive subsurface environment is essential to set up a secondary field measured with inductive EM methods. Electromagnetic methods as a tool for geophysical investigation and exploration is most suited for the detection of water—bearing formation (aquifers) and high – conductive subsurface target such as salt water saturated sediments.

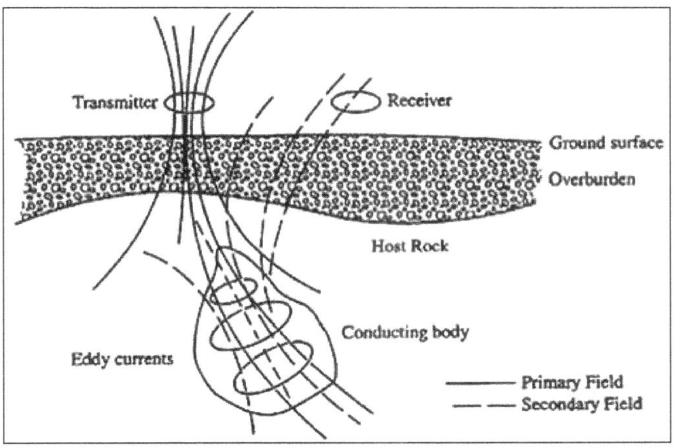

Electromagnetic induction technique.

Instrumentation could take in varying forms; but mainly consist of a source and receiver or receiver units. The source (transmitter) transmits time-varying magnetic fields with the receiver measuring components of the total (primary and secondary) field, magnetic field, sometimes the electric field and the necessary electronic circuitry to process, store and display signals. Data obtained from electromagnetic surveys, like their resistivity counterpart can be collected in profile and sounding mode with their information been presented as maps or pseudo-section to give a better picture of the subsurface. Acquisition, resolution and depth of investigation from this survey are been governed by mostly by conditions of the subsurface and domain of measurement.

EM surveys are divided into two domain system of measurement namely; frequency and time domain system. For frequency domain EM systems, we have the transmitter classed as either high or low frequency transmitters; high transmitter frequencies permits high- resolution investigation of subsurface conductors at near-surface or shallow depths while lower transmitter frequencies allows for deeper depth of investigation at the expense of resolution. This implies that high frequency EM surveys yield better result for near-surface due to high resolution, however if interested in deeper subsurface investigation (low frequency EM surveys) then we have need a way around the low resolution. In the case of time domain system, secondary magnetic field is measured as a function of time, with early–time measurement being suited best for near-surface information while late-time measurement yields results of the deeper subsurface. It is paramount to note that depth of penetration or investigation and resolution is also been governed by coil configuration; while measurements from coil separations are influenced by electrical properties thus the larger coil separation investigates greater depths while smaller coil separation investigates near-surface.

Because Electrical Conductivity is related inversely to Electrical Resistivity, as such discussions relating electrical resistivity to lithology or hydrological properties can be applied in an inverse manner to measurements involving electrical conductivity. Electrical conductivity for example is higher for saturated sediments, clayey materials than for unsaturated sediments and sandy materials respectively. Some examples of investigations involving EM surveys include Sheets and Hendricks, who used EM induction methods to estimate soil water content and McNeill that discussed the relation between electrical conductivity and hydrogeological parameters of porosity and saturation.

Aquifer Characterization – Ground Waves Techniques

Ground Penetrating Radar

Ground Penetrating Radar (GPR) as a geophysical technique is relative new and becoming increasingly popular critically understanding the events of the near-surface or shallow subsurface. Davis and Annan viewed the Ground Penetrating Radar (GPR) as a technique of imaging the subsurface at high resolution using electromagnetic waves transmitted at frequencies between 10 to 1000 MHz. GPR could also be viewed as a non-destructive geophysical technique due to its successful geological applications in urban and sensitive environments. Some of these applications include the subsurface mapping of water table soils and rocks structures (e.g. groundwater channels) at high resolutions. It is similar in principle to seismic reflection profiling in however, propagation of radar waves through the subsurface is controlled by electrical properties at high frequencies.

The GPR survey system is made up of three vital components; a transmitter, a receiver directly connected to the antenna and the control unit. The transmitter radiates EM waves into the subsurface that could be refracted, diffracted or primarily reflected depending on the dielectric permittivity and electrical conductivity nature of the subsurface interfaces encountered. Recorded radar data received after the survey is first been observed, analyzed and interpreted by the aid of inbuilt radar processing software like RADPro, Ekko depending on the system type and make. These data are presented in form of radargrams which could either be presented as 2D or 3D subsurface images depending on the combination of the different axes (x, y and z) involved. Interpreting of radargrams is performed by interface mapping which is quite similar to the technique used in interpreting of seismograms. Here each band within on a radargram is presumably classed and identified as a distinct geological horizon; this would have been correct except for the effects of multiples, interference with previous reflections, noise etc. All this effects on the radargram need to be removed to correctly identify the different geological horizons and geological structures as present within the radargram as such radargram are subjected to varying radar processing operations depending on the aims, objective of the survey been undertaking through the help of inbuilt system radar processing software like RADpro, Pulse Ekko system software etc.

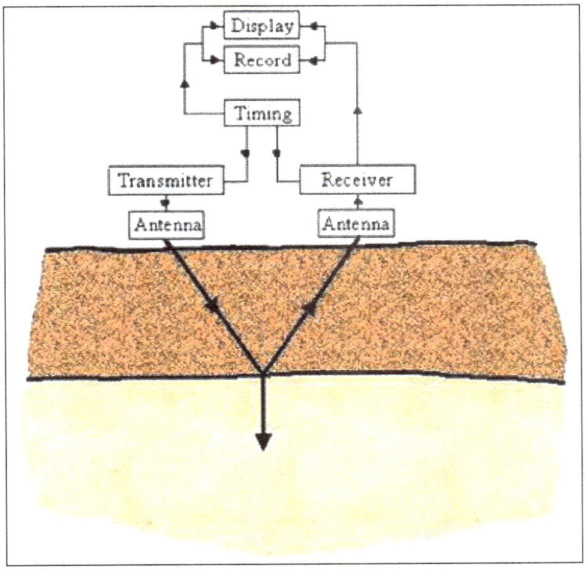

Flow chart for a typical GPR system.

Processing of the radargram could be simplified by processing operations such as dewowing (removal of low frequency components), Gain Control (strengthen weaker events), deconvolution (restores shape of downgoing wave train such that primary events could be recognized more easily), Migration (useful in removing diffraction hyperbolae and restoring dips). The resultant radargram when correlated with the subsurface geology shows varying interfaces, geological structures that might be present. Though GPR has successfully been utilized in unsaturated (non-electrically conductive or highly resistive) and saturated (electrically conductive) environment, however performance is higher in unsaturated (non-conductive) than in saturated (conductive) such as non-expanding clay environment such as at Savannah River Site in South Carolina.

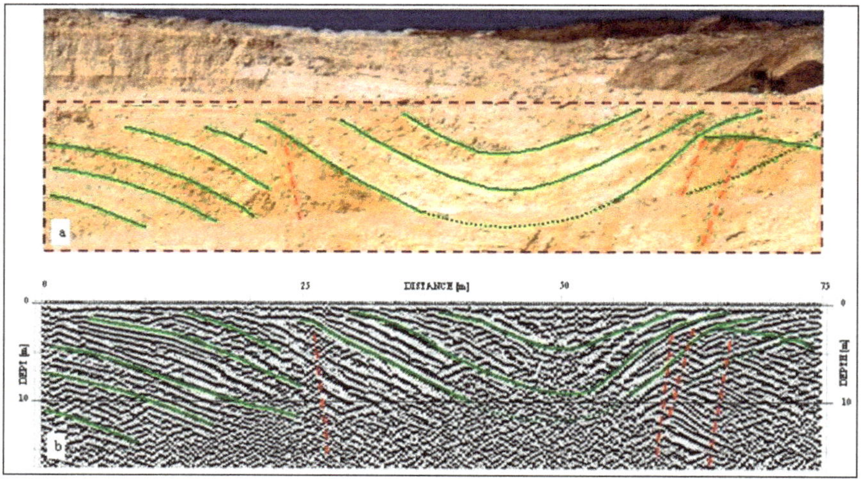

(A) Interpretation of a GPR profile image (B) Interpretation of the prominent stratigraphic units, structures and faults.

The depth of penetration or investigation of GPR survey is function of the frequency of the EM waves or radar waves and nature of the subsurface material been investigated as shown in figure for varying subsurface materials at frequencies ranging between 1 and 500 MHz. If the nature of subsurface material is highly resistive and has low conductivity then we expect a higher depth of penetration however for subsurface materials that are less resistivity and very conductive we expect low depth of penetration. Depth of penetration asides from been dependent on nature of the subsurface material (i.e. resistivity or conductivity nature) is also a function of frequencies which in turn affects resolution of subsurface imagery or radargram. Thus at low frequencies, we expect a greater depth of penetration at the expense of resolution while at high frequencies, we achieve a lower depth of penetration at higher resolution.

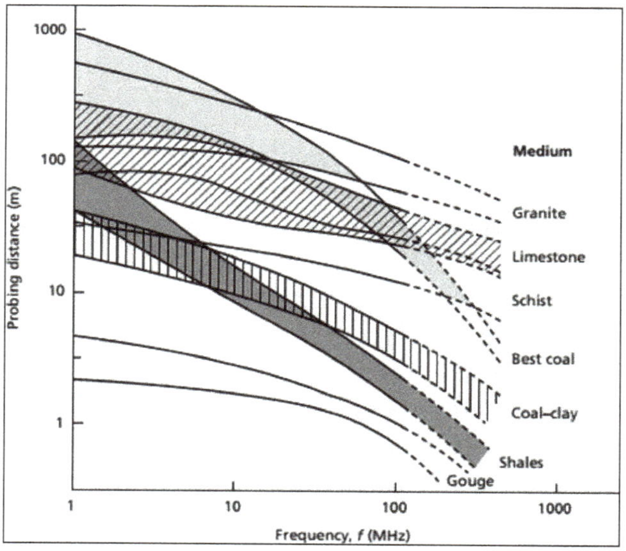

The relationship between probing distance and frequency for different materials.

Ground Penetrating Radar (GPR) data have been successful utilized in the hydrogeological investigations to locate the water table and to delineate shallow, unconsolidated aquifers.

Seismic Techniques

The use of Seismic techniques in subsurface characterization is based on the propagation of elastic waves generated from a seismic controlled source, propagated through the subsurface, boreholes, received by receivers (geophones or hydrophones) and displayed on seismographs (as a combination of waves velocities and attenuation). From these, properties of the subsurface like porosity, hydraulic conductivity, elastic moduli and water saturation which could help us better understand the subsurface could be derived.

Subsurface investigation involving Seismic techniques are categorized into three; Seismic Refraction, Cross-hole transmission (tomography) and Seismic Reflection.

With Seismic Refraction, the incident ray is refracted along the target boundary before returning to the surface. The arrival times gotten from the refracted energy are displayed as function of distance from the source with their interpretation been made manually using simple software or forward modeling techniques. The relationship between arrival times and distances could be used to obtain velocity information directly. Seismic Refraction techniques are the most appropriate for a few shallow (50 m) targets of interest, or where one is interested in identifying gross lateral velocity variations or changes in interface dip. Though Seismic Refraction yields lower resolution than Seismic Reflection and Seismic Cross-hole tomographic, it is however chosen over Reflection as they are inexpensive and help to determining the depth to the water table (buried refractor) and to the top of bedrock, the gross velocity structure, or for locating significant faults. The buried refractor is usually saturated and has a greater velocity than the unsaturated equivalent soil unit and the bedrock surface.

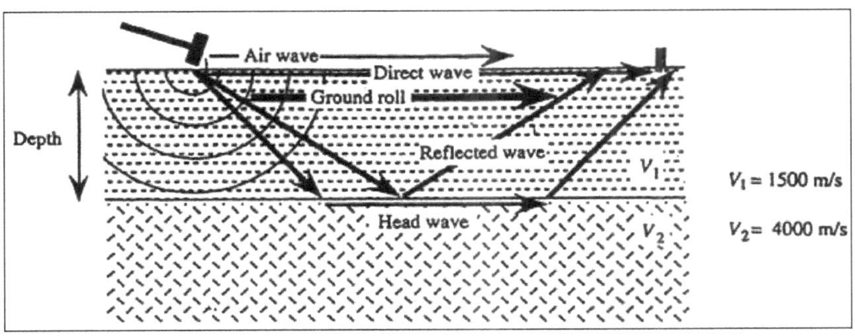

Major ray paths of P-wave energy.

Cross-hole transmission (tomography) data acquisition is possible using several techniques amongst which includes seismic techniques, electrical resistivity, electromagnetic, radar with seismic being the most common. Majority of cross–hole tomographic seismic data have been collected for research however the those collected over extremely high resolution of up to 0.5 m are better suited for site characterization. Figure shows a typical example of seismic cross-hole survey. The multiple sampling of the intra-wellbore area permits very detailed estimation of the velocity structure. As seismic P-wave velocities can be related to lithological and hydrogeological parameters as discussed above, this extremely high resolution method is ideal for detailed stratigraphic and hydraulic characterization of interwell areas.

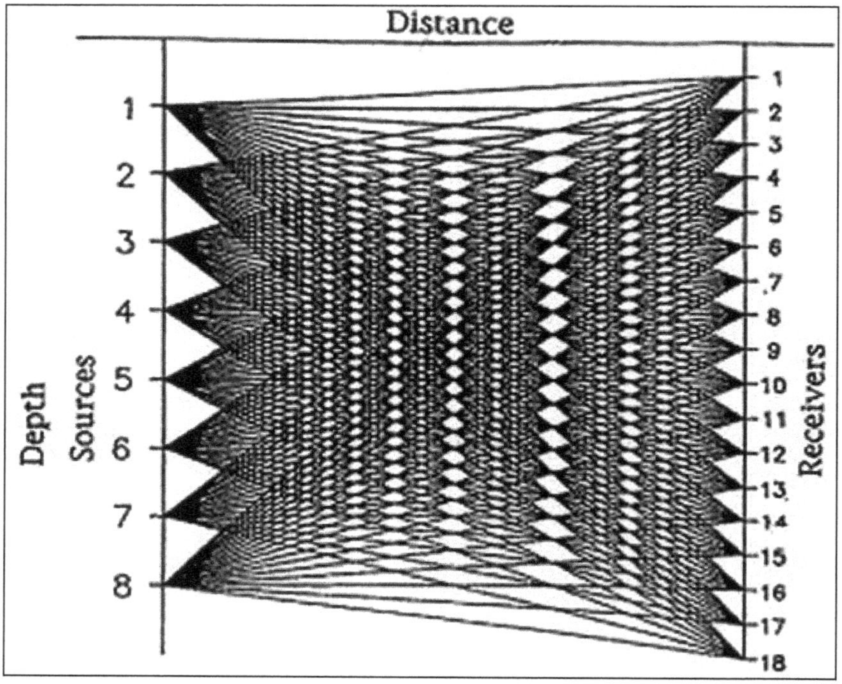

Cross-hole tomography geometry for seismic and radar methods. Sources and receivers are located in separated boreholes, and energy from each source is received by all geophones. Cross-holes acquisition geometries have also been used with electrical resistivity and EM methods.

Water Balance

Water is a dynamic entity, is always in constant motion in the environment and has the ability to change its phase as per the variations in the ambient atmosphere. The hydrologic cycle is a natural cycle that directs and influences the flow of water in the environment and forms a fundamental concept in hydrology. It is a complicated relationship between precipitation and runoff. The water balance concept forms an integral part in hydrological estimates as it helps in understanding the paths and storage of water resources in a system. It helps in examining the hydrologic cycle for any period of time and also estimating the impact of changes in any of the components, i.e., inflows or outflows have on the overall balance of the system. It is an efficient method to study the impacts of anthropogenic activities on the water balance of any subsystem of the hydrologic cycle for any size of area, and at any time period.

Principle of Conservation

The principle of conservation also understood as the principle of conservation of mass states that the mass of a closed system, i.e., a system that restricts transfer of mass and energy, the total mass of the system always remains constant and hence the quantity of 'mass' remains conserved. It is based on the fact that mass can neither be created nor destroyed. This principle finds immense applications in understanding the flows of mass and energy in physical systems operating in the atmosphere. It states that certain tangible physical properties of a system remain conserved in a closed or isolated system. The principle of conservation helps in predicting the physical changes occurring in a system at a macroscopic level without getting in details of the physical and chemical changes occurring at the microscopic level.

Water Budget Equation

The principle of conservation can be effectively applied to the water systems of the atmosphere to comprehensively study the hydrologic cycle. The hydrologic balance is expressed by the equation of continuity which balances the total inflow and outflow of water in a water body. The water balance model for a water system can be conceptually understood by figure.

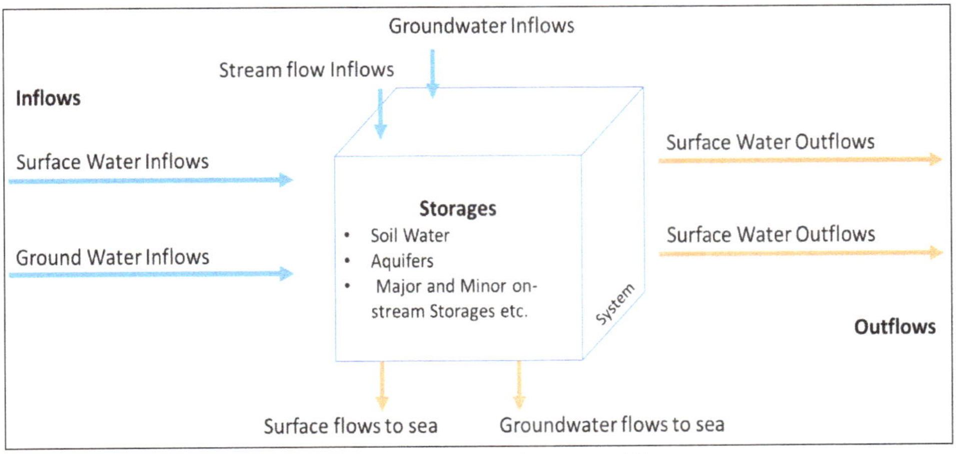

Concept of Water Balance Model.

The water flows represented in the above figure can be expressed in the form of a simple mathematical model. This is the difference in the inflow and outflow of water in a given watershed as the total change in storage as explained in equation:

$$I - O = \Delta S$$

Where, I = Mass inflow; O = Mass outflow; and ΔS = change in mass storage.

The simplified form of water budget equation for a catchment that balances the water inflows and outflows can be expressed by equation:

$$P = Q + E \pm \Delta S$$

Where, P = Precipitation; Q = Runoff; E = Evaporation; and ΔS = change in storage.

This equation shows that water is entering (inflow) in a system in the form of precipitation and leaving (outflow) in terms of evaporation and surface runoff. Therefore water budget equation has important application in the hydrological assessments as it helps the hydrologists to trace the pathways and changes in water storage in a system. The water budget equation is applicable over any space and time scales. The water budget equation can be altered in terms of inflows and outflows to accurately reflect the characteristics of any water system. The water balance equation can be applied at the global as well as a system scale such as lake, stream, river and groundwater by altering the factors for inflows and outflows.

Global Water Balance

Global water balance is the water budget that is prevalent in the atmosphere at the global scale, i.e., including all the continents of the world. The world is known to have ~1386 million cubic kilometers of water. Majority of the water (around 96.5%) present in the world is saline in nature and is stored in the oceans. However, around 1% of the water present on land is also saline. Accounting the total saline water prevailing in the atmosphere, it is estimated that only 35.0 Mm^3 of fresh water is available of which ~68% is contained in a frozen state in the Polar Regions and on mountain tops and glaciers. The distribution of water on the earth surface is presented in table.

Table: Quantities of Water in World.

Sr. No.	Item	Area (Mkm^2)	Volume (Mkm^3)	Percent total water	Percent fresh water
1	Oceans	361.3	1338	96.5	-
2	Ground Water: fresh saline	134.8 134.8	10.530 12.870	0.76 0.93	30.1
3	Soil moisture	82.0	0.0165	0.0012	0.05
4	Polar ice	16	24.0235	1.7	68.6
5	Other ice and snow	0.3	0.3406	0.025	1.0
6	Lake: fresh saline	1.2 0.8	0.0910 0.0854	0.007 0.006	0.26 -
7	Marshes	2.7	0.01147	0.0008	0.03

8	Rivers	148.8	0.00212	0.0002	0.006
9	Biological water	510	0.00112	0.0001	0.003
10	Atmospheric water:	510	0.01290	0.001	0.04
Total	All kinds of water Fresh Water	510 148.8	1386 135	100 2.5	100

The water present in various forms across the world maintains a balance and the overall quantity of water remains conserved. The global annual water balance is presented in table below. The amount of water that evaporates over the oceans is estimated to be 9% higher than the amount of water that is restored back in the oceans as precipitation. However, it is observed that the amount of precipitation received by land is higher in comparison to that of oceans. Thus, the differential of water amounting to 0.047 Mm^3 is replaced by the surface runoff and groundwater outflow to the oceans in order to maintain the balance of water in the system. This represents the overall balance in the water system that is maintained by different forms of water bodies and different processes occurring in the environment.

Table: Global Annual Water Balance.

Sr. No.	Item	Ocean	Land
1	Area (M km^2)	361.30	148.8
2	Precipitation (km^3/year) (mm/year)	458,000 1270	119,000 800
3	Evaporation (km^3/year) (mm/year)	505,000 1400	72,000 484
4	Runoff to oceans: Rivers (km^3/year) Groundwater Total runoff (km^3/year) (mm/year)		44,700 2,200 47,000 316

System Water Balance

Water Balance of a Lake

A lake is a slow moving water body that is formed due to the accumulation of water in surface depressions. The lakes have two predominant motions of water, i.e., vertical movement also defined as the movement in columns (surface to bottom) and horizontal movement also defined as the movement through the length of the lakes (inflow to outflow). The water balance of a lake takes into account all the movements undertaking within this system and is determined using the mathematical relation presented in equation:

$$Q_{in} - Q_{out} = \frac{dV}{dt} = A\frac{dh}{dt}$$

Where, Q_{in} = Input flux; Q_{out} = Output flux; V = Volume of lake (m^3); dV/dt = Rate of volume change with time; h = depth of water (m); A = Area of lake (m^2).

In order to estimate the total storage, it is very important to identify the pathways of inflow and outflow in a lake. Figures provide a typical example of the key pathways of movement of water in a lake.

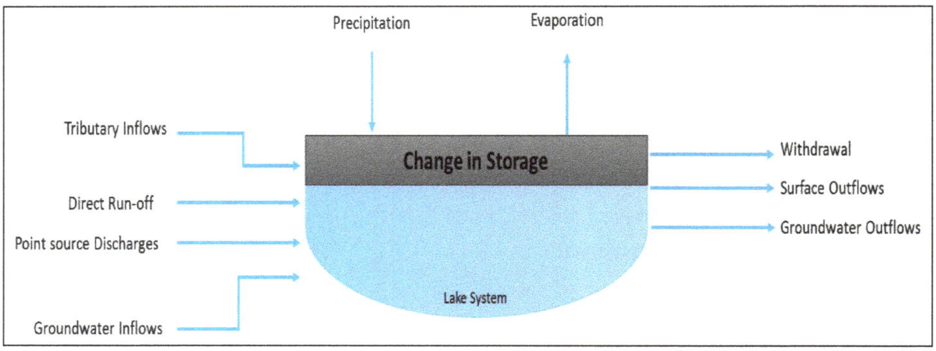

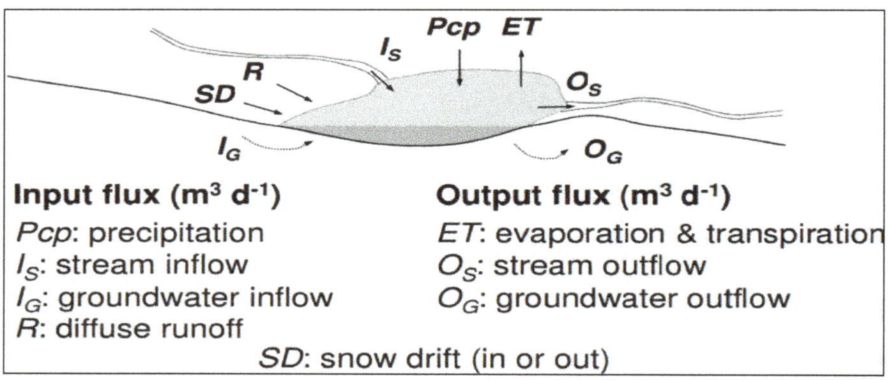

Water Flow Pathways in a Lake System.

The key components of the water balance equation of a lake are precipitation, evaporation, evapotranspiration, surface runoff and groundwater flow. The water balance equation for a lake thus gets modified as per the parameters defines in the above figures and can be represented by equation:

$$P - R - G - E - T = \Delta S$$

Where,

- P = Precipitation,
- R = Surface runoff,

- G = Net ground water flow out of the catchment,
- E = Evaporation,
- T = Transpiration,
- ΔS = Change in water storage.

Mathematical Forms of Water Balance for Different Zones

Water Balance for Unsaturated Zones

The water balance of an unsaturated zone is predominantly governed by the sub-surface flows that determine the soil moisture. Thus, the soil-moisture budget can be established for an unsaturated zone on the concept of water balance equation by measuring the inputs through precipitation and its distribution in the plants, evaporation and soil moisture. Figure presents a pictorial presentation of an unsaturated zone, i.e., the zone where soil-water balance is prevalent that determines the pathways and storage of water.

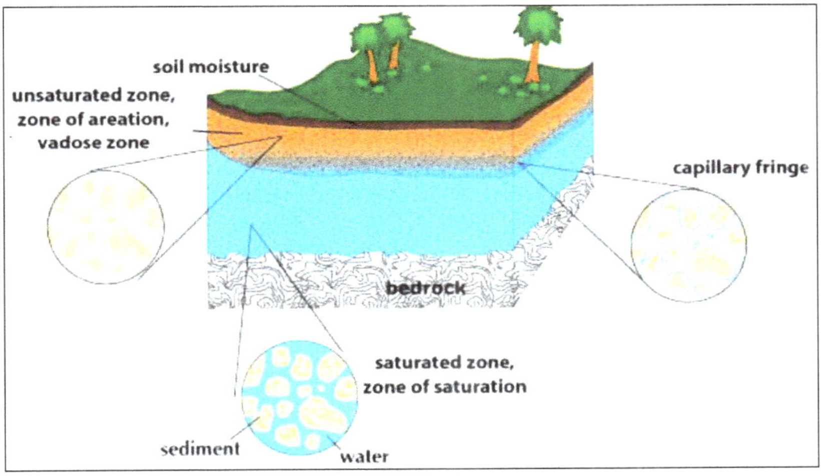

Soil-Water Balance.

Equation presents the mathematical equation that can be used to estimate the water balance of an unsaturated zone,

$$I - E + G - R = \frac{\Delta W}{\Delta t}$$

Where,

- I = Infiltration rate of water into unsaturated zone (mm/d).
- E = Evapotranspiration rate of water from unsaturated zone (mm/d).
- G = Rate of capillary rise from the saturated zone (mm/d).

- R = Rate of percolation to the saturated zone (mm/d).
- ΔW = Change in soil water storage in the unsaturated zone (mm).
- Δt = computation interval of time (d).

Water Balance for Land Surfaces

The mathematical equation for estimating the water balance for land surfaces is described in equation,

$$I = P - E_l + 1000\frac{Q_i - Q_o}{A} - \frac{\Delta W_S}{\Delta t}$$

Where,

- I = Rate of infiltration into the unsaturated zone (mm/d).
- P = Precipitation for time interval Δt (mm).
- E_l= Evaporation from the land surface (mm/d).
- Q_i = Lateral inflow of surface water into the catchment area (A) (m^3/d).
- Q_o = Lateral outflow of surface water into the catchment area (A) (m^3/d).
- A = Water balance area (A^2).
- ΔW_s = Change in surface water storage in the unsaturated zone (mm).

Ground Water Balance

The groundwater zone is a complex system comprising of three regions which accommodate the predominant flows of water, i.e., zone of aeration, zone of saturation and the water table. The three regions can be briefly understood as under:

- Zone of Aeration: This is the uppermost zone which is nearest to the surface made of soil and rock. This is the zone through which the excess surface water flows and the gaps between the soils are filled with both air and water.
- Water Table: Water table is the region that separates the zone of aeration and the zone of saturation. It is primarily known as the region prevalent on top of the saturation zone.
- Zone of Saturation: The zone underneath the zone of aeration is known as the saturation zone. This zone is characterized by rocks and pores that are filled up to saturation level by subsurface water.

A diagrammatic representation of the various zones in a groundwater system present below the land surface layer is provided in figure.

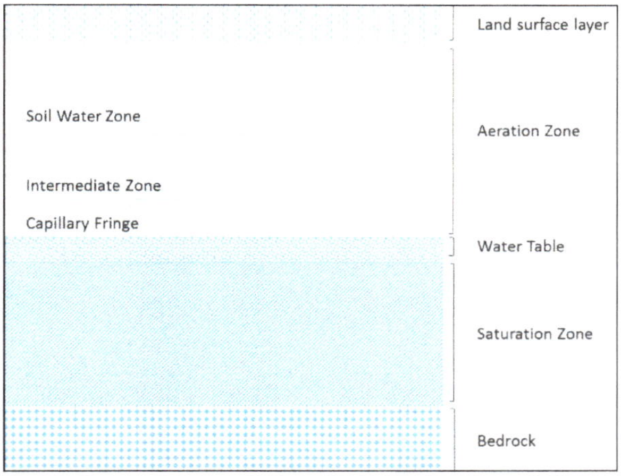

Zones of Subsurface water.

The water balance of a saturated zone is used to represent the ground water balance equation. The mathematical form for the groundwater balance can be expressed using equation,

$$R - G + 1000\frac{Q_{gi} - Q_{go}}{A} = \mu\frac{\Delta h}{\Delta t}$$

Where, Q_{gi} = Total rate of groundwater inflow in the shallow unconfined aquifer (m^3/d) as represented by equation,

$$Q_{gi} = Q_{gih} + Q_{giv}$$

Where,

- Q_{gih} = Rate of horizontal groundwater inflow into the shallow unconfined aquifer (m^3/d).
- Q_{giv} = Rate of vertical groundwater inflow from the deep confined aquifer into the shallow unconfined aquifer (m^3/d).
- Q_{go} = Total rate of groundwater outflow from the shallow unconfined aquifer (m3 /d) as represented by equation.

  $$Q_{go} = Q_{goh} + Q_{gov}$$

Where,

- Q_{goh} = Rate of horizontal groundwater outflow from the deep confined aquifer into the shallow unconfined aquifer (m^3/d).
- Q_{gov} = Rate of vertical groundwater outflow from the deep confined aquifer into the shallow unconfined aquifer (m^3/d).
- μ = Specific yield.

- Δh = Rise or fall of water table during the computation interval (mm).

Apart from equation $R - G + 1000\frac{Q_{gi} - Q_{go}}{A} = \mu\frac{\Delta h}{\Delta t}$, a detailed equation for groundwater balance is presented in equation below, which represents a balance in the total inflows and total outflows to the groundwater system. The inflows are majorly determined in the form of groundwater recharge while the outflows are primarily in the form of evapotranspiration and water drafts from the groundwater.

$$R_r + R_c + R_i + R_t + S_i + I_g = E_t + T_p + S_e + O_g + \Delta S$$

Where,

- R_r = Recharge from rainfall.
- R_c = Recharge from canal seepage.
- R_i = Recharge from field irrigation.
- R_t = Recharge from tanks.
- I_g = Inflow from other basins.
- E_t = Evapotranspiration from groundwater.
- T_p = Draft from groundwater.
- S_e = Effluent seepage to rivers.
- O_g = Outflow to other basins.
- ΔS = Change in storage.

Integrated Water Balance Equation

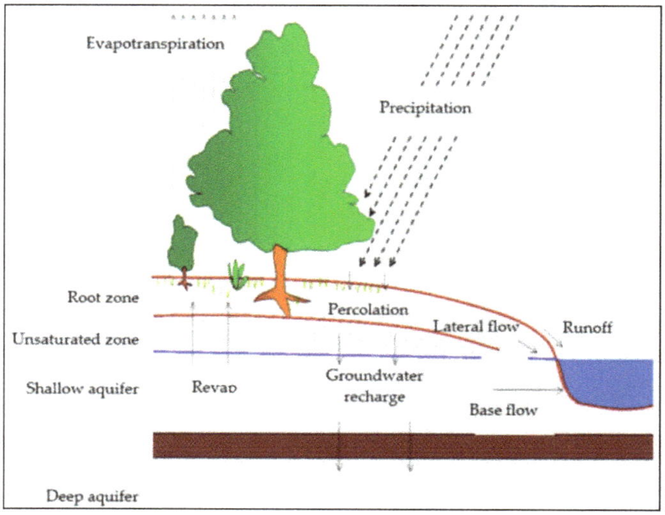

Integrated Movement of Water land surfaces, unsaturated zones and groundwater.

The integrated water balance equation combines all the three water balances operating at the land surfaces, unsaturated zones and groundwater. The integrated water balance equation helps in describing the entire hydrologic cycle that accounts for all the major inflows and outflows of water in the atmosphere. It provides a holistic picture of the variations that occur in the pathways of water flows and total storage in the atmosphere by simultaneously analyzing the water balances occurring at smaller scales of the hydrologic cycle. Figure provides a description of the integrated water movement occurring in the atmosphere.

Equation represents the mathematical form of integrated water balance equation,

$$P - E_o - E + 1000\frac{Q_{si} - Q_{so}}{A} + 1000\frac{Q_{gi} - Q_{go}}{A} = \frac{\Delta W_u}{\Delta t} + \frac{\Delta W_\varepsilon}{\Delta t} + \mu\frac{\Delta h}{\Delta t}$$

Where, P = Precipitation for time interval Δt (mm); E_o = Rate of evapotranspiration from the unsaturated zone (mm/d).

Water Sustainability in a Dynamic World

Water is the largest limited natural resource which is vital for survival of all living beings. Floods, droughts, and famines brought about by climate changes have been noted to occur with more frequency in the recent years. Therefore, the need for the adoption of sustainable methods toward water use and management is critical in the present-day scenario. In addition, there is also an urgent need to develop policies and make smart investment decisions to promote water sustainability in the light of climate change.

Water Footprint

Increase in standard of living around the world has also brought with it an increase in the demand for water-intensive goods further impacting the already limited fresh water resources. Water footprint is the term used to calculate the amount of water pertaining to the production of a commodity. The water footprint can either be classified as green, blue, or grey depending on the source of water used for producing a product. While green refers to the amount of rainwater, blue refers to the amount of groundwater used to produce a product. On the other hand, grey water footprint refers to the amount of fresh water that will be used to dilute an aqueous solution consisting of pollutants to bring it down to the desirable level. It is critical that countries take steps to keep the water footprint at the lowest and promote sustainable management of water resources as the implications of irresponsible water usage can be global in nature.

Water-Stressed Cities

Studies from The Nature Conservancy report indicate that one in four major cities is water stressed, and the demands for water will double in the next three decades. The Water Scarce Cities (WSC) Initiative was initiated by the World Bank to promote water security. This was promoted by increasing awareness among people and nations, facilitating collaborations and dialogs between stakeholders, and by facilitating and providing technical assistance and logistical support in countries. In the present situation of water scarcity, it is critical to rethink related aspects of urban water management in the light of rainwater harvesting, water use, and wastewater treatment since all these sectors need to be handled together.

Revival and Rehabilitation of Waterbodies

Water management for a sustainable future also includes rejuvenation and revival of polluted waterbodies. Revival of the waterbody promotes the conservation of the ecology of the surrounding area. Rehabilitation of waterbodies is first brought about by cleaning/removal of the sludge settled, followed by methodical ways to stem the flow of effluents from industries or raw sewage into the waterbody. Recently, there has been an increased interest in exploring eco-friendly approaches in treatment of waterbodies. In the same vein, the potential of floating wetlands in promoting water purification has been widely studied. Floating wetlands often consist of hormonally treated plants with synergistic bacterial colonies in the rhizosphere region which have the ability to adsorb harmful trace contaminants in the waterbody.

Promoting Clean Water and Cities

Proper disposal practices with regard to hazardous chemicals (e.g. fertilizers) can go a long way in reducing the load of nutrients reaching the waterbody. Excessive levels of nitrogen and phosphorus in waterbodies (from surface run-off from agricultural fields) can promote eutrophication of water bodies leading to the destruction of the fresh water ecosystem. Maintaining a low level of water footprint and judicious water reuse strategies can significantly reduce the water consumption and reduce the water stress. Water reuse strategies such as a grey water pool system can be undertaken at a larger scale in communities to promote sustainable water management. Adoption of techniques like rainwater harvesting to recharge the aquifer as well as use it as a source of water can be promoted extensively thereby minimizing run-off of precious rainwater into open drains.

References

- Different-types-of-water-sources: h2ocoolers.com, Retrieved 14, August 2020
- Aquifer-classification-and-characterization, aquifers-matrix-and-fluids: intechopen.com, Retrieved 21, April 2020
- Introductory-chapter-water-sustainability-in-a-dynamic-world, water-and-sustainability: intechopen.com, Retrieved 05, July 2020

2

Water Resources Modeling and Management

There are many types of modeling approaches to provide a better understanding of alternative water resource designs and management plans. Some of these modeling approaches are catchment modeling, river basin modeling and rainfall-runoff modeling. All these types of modeling in water resources have been carefully analyzed in this chapter.

Modeling provides a way, perhaps the principal way, of predicting future behavior, at least in a statistical sense, of existing or proposed water resource systems. The past 30 years have witnessed major advances in our abilities to model the engineering, economic, ecologic, hydrologic, and sometimes even the institutional or political aspects of large complex multipurpose water resource systems. Applications of models to real systems have improved our understanding of such systems, and hence have often contributed to improved system design, management, and operation. Evaluating the applications of numerous types of models has also taught us how limited our modeling skills remain.

Water resource systems planning, management, and operation is far more complex than what analysts have been, or perhaps even will be, able to model and solve. The reason is not simply any computational limitations on the number of model variables, constraints, subroutines, or executable statements in those subroutines. Rather it is because we do not understand sufficiently well the multiple interdependent physical, biochemical, ecological, social, legal and political (human) processes governing the behavior of such water resource systems. These processes are affected by uncertainties in what we manage, e.g., water quantities and qualities and water demands, as well as in the models we use to describe them. Uncertainties also result from the unpredictable actions of multiple individuals and institutions that are impacted by what they get or do not get from the management and operation of such systems, as well as by other events having nothing to do with water.

The development and application of models, or the practice of modeling, should be preceded by the recognition of what can and cannot be achieved from the use of such models. Models of real-world systems are always going to be simplified representations of those systems. What features of the actual system are incorporated into a model, and what features are not, depends in part on what the modeler thinks is important with respect to the issues. How well this is done depends on the skill of the modeler, the time and money available, and, perhaps most importantly, the modeler's understanding of the real system and decision-making processes. Developing models is an art, requiring knowledge of the system being modeled, the client's objectives, goals and information needs, and some analytical and programming skills. Models are always based on numerous assumptions, and some of these may be at issue. Applying these approximations of reality in a way that contributes to everyone's improved understanding and eventually to a good decision clearly requires considerable modeling and communication skills as well as a little bit of experience.

Water resource planners and managers must accept the fact that decisions may not be influenced by their planning and management model results. To know, for example, that cloud seeding may, on average, reduce the strength of hurricanes over a large region does not mean that such cloud-seeding activities will or should be undertaken. Managers or operators may know that not everyone may benefit and those who may lose will likely scream louder than those who may gain. Hence, decision-makers may feel safer in inaction than action. There is a strong feeling in many cultures and legal systems that failure to act (nonfeasance) is considered more acceptable than acts that fail (misfeasance or malfeasance). We all feel greater responsibility for what we do than for what we do not do. Yet our aversion to risk should not deter us from addressing such sensitive issues in our models. After all, our modeling efforts should be driven by the need for information and improved understanding. It is an improved understanding (not improved models per se) that may eventually lead to improved system design, management, and/or operation. Models used to aid water resource planners and managers are not intended to be, and rarely are (if ever), adequate to replace their judgment.

Challenges of Planners and Managers

Planners and managers of water resource systems are those who are responsible for solving our particular water-related problems or meeting our special water resource needs. When they fail, the public lets them know. What makes their life particularly challenging is that the public has different needs and expectations. Furthermore, water resource institutions in which planners and manager's work (or hire consultants to work for them) are like most institutions these days. They must do what they can with limited financial and human resources. Their clients are all of us who use water, or at least all of us who are impacted by the decisions they must make. The overall objective of these institutions is to provide a service, whether it is water supply, water quality, hydropower, flood control and protection, navigation, recreation, wildlife preservation, or some combination of these or other purposes. Furthermore they are expected to do this at a price people are willing to pay. Meeting these goals is not always easy, or even possible.

There are rarely simple technical measures or procedures available to ensure that a solution to any particular set of water resource management problems can be achieved. Furthermore, everyone who has had any introduction to water resources planning and management knows one cannot design or operate a water resource system without making compromises. These compromises are over competing purposes (such as hydropower and flood control) or competing objectives (such as who benefits and who pays, and how much and where and when). After analysts identify possible ways of achieving various goals and objectives and some of their economic, environmental, ecological and social impacts, it is the planners and managers who have the more difficult job. They must work with and influence everyone who will have a role or stake in deciding what to do and/or in implementing the decision.

Planning and managing involves not only decision-making, but also developing among all interested and influential individuals an understanding and consensus that legitimizes the decisions and enhances their successful implementation. Planning and managing are processes that take place in a social environment. These processes involve leadership and communication among people and institutions. Leadership and communication skills are learned from experience, not from computers or models.

Moving an organization or institution into action to achieve specific goals involves a number of activities, including goal-setting, debating, coordinating, motivating, deciding, implementing, and monitoring. Many of these must be done simultaneously and continuously, especially as conditions (goals, water supplies, water demands, and finances) change over time. These activities create a number of challenges that are relevant to modelers or analysts. Some include how to:

- Identify creative alternatives for solving problems.
- Make decisions and implement them given differences in opinions, social values, and objectives.
- Find out what each interest group wants to know in order to reach an understanding of the issues and a consensus on what to do.
- Develop and use models and present their results so that everyone is able to reach a common understanding and agreement that is consistent with their individual values.

In addressing these needs or challenges, planners and managers must consider the relevant:

- Legal rules and regulations.
- History of previous decisions.
- Preferences of important actors and interest groups.
- Probable reactions of those affected by any decision.
- Relative importance of various issues being addressed.
- Applicable science, engineering, and economics—the technical aspects of their work.

Challenges of Modeling

To engage in a successful water resource systems study, the modeler must possess not only the requisite mathematical and systems methodology skills, but also an understanding of the environmental engineering, economic, political, cultural, and social aspects of water resources planning problems. For example, to study the impact of a large land development plan, the analyst should be able to predict how the proposed plan would affect runoff and, in turn, the quantity and quality of surface waters and ground waters, and how the development would affect flood flows and conversely, how flood flows would affect the planned development. Add to this the prediction of the increasingly important ecological impacts resulting from the land uses and water quantity and quality regimes we create. To do this the analysts must have an understanding of the biological, chemical, and physical and even social processes that are involved in water resources management.

A reasonable knowledge of economic theory is just as important as an understanding of hydraulic, hydrologic, ecologic and environmental engineering disciplines. Economics has always had, and will continue to have, a significant role in the planning of water resources investments. It is obvious that the results of most water resources management decisions have a direct impact on people and their relationships. Hence inputs from those having a knowledge of law, regional planning, and political science are also needed during the comprehensive planning of water resource systems, especially during the development and evaluation of the results of various planning models.

Some of the early water resource systems studies were often undertaken with a naive view of the appropriate role and impact of models and modelers in the policymaking process. The policymaker could foresee the need to make a decision. He or she would ask the systems group to study the problem. They would then model it, identifying feasible solutions and their consequences, and recommend one or at most a few solutions. The policymaker, after waiting patiently for these recommendations, would then make a yes or no decision. Experience to date suggests the following:

- A final solution to a water resources planning problem rarely exists; plans and projects are dynamic and change and evolve over time as facilities are added and modified and the uses and demands placed on the facilities change.
- For every major decision there are many minor decisions, made by different agencies or management organizations responsible for different aspects of a project; plans evolve.
- The time normally available to study a water resources problem is shorter than the time necessary for an adequate state-of-the-art mathematical modeling study, or if there is sufficient time, the objectives of the original study will have significantly shifted by the time the study is completed.

This experience emphasizes not only some of the limitations and difficulties that any water resource systems study may encounter, but more important, it emphasizes the need for constant communication among the analysts, engineers responsible for the systems operations, and policymakers. The success or failure of many past water resource studies is in a large part attributable to the efforts expended or not expended in ensuring adequate and meaningful communication—communication among systems planners, professional engineers responsible for system operation and design, and public officials responsible for major decisions and setting general policies.

It is these engineers and public officials, after all, who need the information that can be derived from various models and analyses, and they must have it in a form useful and meaningful to them. At the beginning of any study, objectives are usually poorly defined. As more is learned about what can be achieved, people are better able to identify what they want to achieve. Close communication among analysts, engineers, and public officials throughout the modeling process is essential if systems studies are to make their greatest contribution to the planning process. Furthermore, those who will use models, and present the information derived from models to those responsible for making decisions, must be intimately involved with model development, solution, and analysis. Only then can they appreciate the assumptions upon which any particular model is based, and hence adequately evaluate the reliability of the results.

Any water resource systems study that involves only outside consultants, and minimal communication between consultants and planners within a responsible management agency or involved stakeholders, is unlikely to succeed in having a significant impact on the planning process. Models that are useful are alive, constantly being modified and applied by those groups which are involved in plan preparation, evaluation, and implementation.

Catchment Modeling

A catchment is an integrated unit of land, soil and water which provides for society, economy and ecosystem. The catchment is an essential unit which must be healthy for a healthy, protected and sustainable environment. A catchment can typically have industries, cities, towns, as many catchments are no longer are purely agricultural, pasture land or covered completely with forests. Similarly, many catchments support lakes, wetlands, are dependent on ground water completely, have many streams, or are a part of a major river system.

Each catchment has specific water resources which have been harnessed for agriculture, industry, domestic use, power generation, drinking water, recreation. This in many regions of the world has led to excessive withdrawals from surface water and ground water so that the current water demands cannot be met, and the diverse ecosystems have started showing signs of degradation. Land is degraded due to excessive use of fertilizers, with intensive agricultural practices, heavy nutrient land or suffering from soil erosion. All this needs the catchment characteristics to be studied in detail and holistically in order to provide support to decision makers. Some of the typical concerns in catchments now due to the excessive use of resources are as follows:

- Excessive withdrawals from surface and ground water.
- Construction of storage reservoirs, ponds at farm level, leading to changes in water received by downstream.
- Deforestation leading to excessive erosion and changes in sediment flow regime in streams.
- Pollution from agriculture and industry practices.
- Eutrophication in lakes and ponds.
- Excessive fish harvesting.
- Reduction in flood plain zones due to construction activities.
- Plantation of crops not native to the catchment, leading to undesirable changes in the catchment.

In addition, forest fires, floods event and droughts further affect the security of available water resources of the catchment. The impacts of all the mentioned factors above are felt on the environment, society and economy. The idea today is to manage the catchment in such a way so that it continues to be in good health for future generations too or is sustainable in the future too. Sharing water resources between competing uses in a sustainable manner is a challenge which has to be addressed. The issues related to water management are multi-dimensional and multidisciplinary, thus many factors need to be considered, so we use the framework of system analysis or modeling to understand the challenges and provide solutions, be it on catchment scale or river basin scale.

The systems analysis or modeling has a demarcated boundary within which the functions take place. Every model has inputs, goes through a connection (systems) and outputs are revealed. Many times the entire catchment cannot be understood with the help of one model only, many subsystems or smaller models (related to a few aspects only) are used which are then collated to understand the complete catchment.

A model typically is based on the following:

- Identification of clear objectives.
- Determining the measurable components in the model.
- Alternative courses available to answer the objectives.
- Comparative evaluation of all alternatives to reach the objectives.

To give an example, input can be climate variables, land use, water use, and population while outputs can be stream discharge, erosion, sediment flow in streams. The alternative systems available to reach these outputs through the given inputs are what are explored in the various modeling systems.

A typical model has the following properties:

- A model must have a clear relationship between input and output or relationship between inflow and outflow.
- Identify and address system constraints like effects on the ecosystem.
- Prioritize between short term and long-term objectives.
- Provision for including stakeholders, so that social, economic and environment are addressed.

A model is prepared as it is easier to work with a model than the real system, at the same time it is a representation of the bigger actual system. What parts of the real system is being modeled depends on the sub system that we are expecting to study in detail. The most common model used in the water balance model for a catchment. The water balance equation is as follows: Inflow (I)-Outflow (O) = change in storage (ΔS). This basic equation is modeled at the catchment scale.

This is the basic hydrological model, where the difference between water inflow and water outflow from the catchment is the change in storage. It represents the continuity equation which is the basis for most hydrological models. The input for the system may be the stream inflow, the stream outflow, while the change in lake level is modeled using the inflow and outflow characteristics to understand the change in lake storage or storage in the catchment in general if the same equation is modeled at the catchment outlet. Some of the common catchment models are on:

- Rainfall runoff relationships (surface water).
- Flow of water in vadose-saturated groundwater zone (sub surface water).
- Surface water-groundwater interactions.
- River basin modeling.
- Models relating water quantity and quality, their interrelationships within the catchment, relationship with the ecosystem and the community.
- These models also represent increasing complexity as the rainfall runoff model has few variables while the model relating quantity to quality may have many variables.

Necessity of Modeling

Many anthropogenic activities can interfere or change the hydrological cycle. Thus, in order to understand the entire catchment holistically, integration of knowledge from within the water management field to across disciplines such as economic, ecological, social factors are considered while analyzing a system. Models help in understanding the system and if a proper interface is developed then through graphs or otherwise scientific results may be communicated to all stakeholders.

Models help in optimization of resources and forecasting and evaluation of resources. A simple rainfall runoff relationship can be established using existing observation, once the model is established; it can be used to predict runoff from different rainfall scenarios. It can also help in optimizing water allocation to different water demands using the quantified runoff.

Classification of Modeling Approaches

They are classified into three types:

- The model may be empirical, conceptual or based on physical phenomena.
- Models may be stochastic or deterministic depending on how inputs are taken in the model.
- Models may follow lumped or distributed approach.

Empirical models are the simplest models, in which the relationship between variables is based on observed data alone. For example, Runoff = a X $(\text{rainfall})^b$ is of an empirical type, which is also called a black box model. This is because only a mathematical relationship is established between the two variables without going into the physical processes involving the catchment. The parameters a and b are derived from regression analysis of the measured rainfall runoff data only. These models though may be accurate for a catchment, may not be replicated for other catchments, or extrapolated for other catchments. These models require data for few variables only and they are region specific.

The next in complexities are conceptual models, also called perceptual models. They use the actual physical condition of the catchment to develop the modeling framework. The hydrological cycle of an area is a conceptual model. The conceptual model of the catchment conceptualizes the many water storage capacity structures and uses mathematical functions to understand the linkages between upstream and downstream surface and sub-surface flows. The conceptual model can be lumped, where all the storage capacity is represented as one unit or semi distributed where the catchment is divided into sub catchment storages which are then linked.

The next categories of models are physically based, or process based models. They are more complicated that the other two categories and use more variables, more parameters to explain the system. MIKE, SHE for surface water flows and MODFLOW for ground water flow are computer based algorithms which are physically based models.

They solve mathematical equations to understand physical processes like conservation of mass or momentum.

The next category of models under discussion is deterministic and stochastic models. The deterministic models are more common, where a single set of input variables result in a single output being determined. In stochastic modeling the variables are represented as statistical distributions, so that a range of output values are generated each derived from a certain set of variable inputs having certain probability of occurrence.

Event Based or Continuous Modeling

When a single storm event is analyzed may be lasting for a few hours or few days, produces a single runoff for the event, it is called event modeling. The other form of modeling is continuous catchment modeling in which surface water balance, subsurface water balance and ground water balance is continuously modeled.

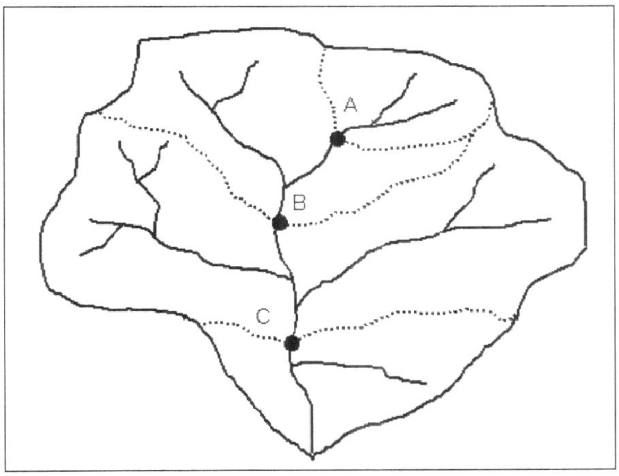

Sub catchment boundaries at A, B and C.

Process based models are the most complicated, these models are based on solving mathematical equations describing fundamental principles like the law of conservation of mass or momentum. Often, they assume distributed modeling approach, as in the catchment is divided into many smaller segments or grids as shown in figure. As the units of data become finer, often they become more complicated and uncertainty in results obtained increases.

Geographic Information System (GIS) in Modeling

GIS allows visualization of the catchment characteristics as free data sets from internet are available for land use, digital elevation, soils and climate variables. So, using GIS tools, studying spatial characteristics of the catchment is becoming more common and therefore hydrological modeling is seeing prolific use of GIS. For example, terrain data sets can be used to determine slope of a catchment which is used as input in hydrological modeling. GIS data uses ground surveys, digitize existing maps, recorded aerial photography, satellite imaging data or a combination of all these. The digital elevation maps help in delineating catchment boundaries and the catchment stream network. Thus, hydrological assessment of catchments is using GIS more and more useful now. There are many hydrological modeling packages using GIS interface.

Modeling Complexity

There is no hard and fast rule that the model must be either empirical or conceptual or process based, it can be a hybrid of all these. It is not always necessary that the most complex model gives the best solutions, the balance between data availability and model complexity gives good prediction. The simplest models are lumped in both space and time while more complex models are distributed in space and time. There are models which give an extremely good fit to data to let us say stream flow, yet at sub catchment level the model does not give representative data as the input parameters needed to represent the catchment a finer scale may be different. Predictive performance of a model is governed not just by model complexity, as the figure below shows model B is far more complex than Model A yet, the predictive performance is not that much better than model A. Optimum data availability and use with model complexity gives optimum predictive performance. To explain modeling complexity, typically flood simulation processes can be used.

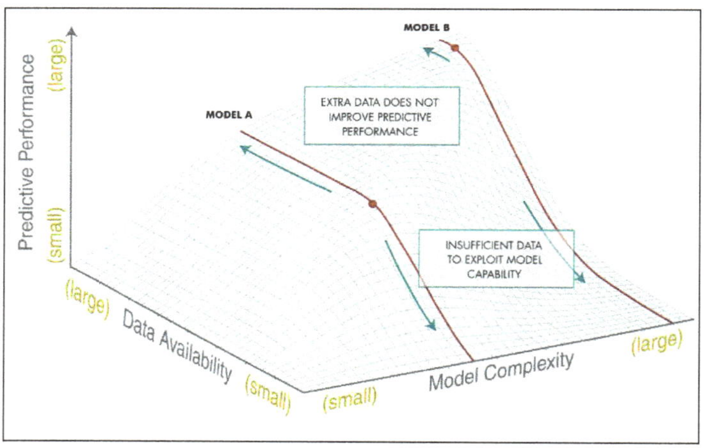

Relationship between complexity of model, availability of data and predictive capacity of the model.

Deterministic Modeling

A flood event is generated with the help of storm hydrograph for the given design rainfall event. Traditionally the temporal rainfall distribution is studied from historical records; the extreme events are chosen for records, volume of runoff generated from such events is determined. However, in these the variability of storm pattern remains challenge which is not addressed. The parameters chosen for such models are the rainfall depth with time, losses, idea about the temporal and spatial pattern of rainfall. All these parameters are fed in the catchment model to generate the flood peak with time.

Probabilistic Modeling

For this kind of modeling, sample stochastic rainfall events are chosen, then a flood frequency curve is generated using storm hydrograph. With Monte Carlo simulations, produces likely flood events hence, allows incorporating uncertainty in flood predictions.

Steps in Modeling

- Establishment of the model,
- Calibration,
- Validation,
- Outcome.

Once the model is established, and then comes the calibration stage. In the calibration stage it is important to understand the control parameters. These parameters are different for gauged and ungauged catchments. In case of gauged catchments both rainfall and runoff data are available, while in case of ungauged catchments, rainfall data may be available, but runoff data is not available. Thus, control parameters are very difficult to establish in case of ungauged catchments. The control parameters in case of ungauged catchments are established based on physical characteristics which are like a gauged catchment having similar characteristics or from regional relationships, if these are already established. For calibration the control parameters must be modified so that the simulated conditions match with the measured results. The calibrated control parameters for one series of storm events may not be valid for a second set of storm events. However, the system needs to be robust if extrapolation is to be allowed.

In the validation step model from a separate time is used to understand the catchment system response. This helps us to assess the accuracy and uncertainty of predictions. One of the most commonly practiced validation process in catchment modeling is called split sample procedure. The data is divided into two groups, one set is used for calibration, using that calibrated model real time data is simulated with the adopted parameters. The next set of data is then used for validating the model. If the data set is large then the data is split in three ways, it is advisable to finally get the parameter values to be used in practice by performing a calibration by using the whole data sample, to reduce the uncertainty of the parameter estimates as much as possible.

Expert knowledge is used to estimate parameter values, and then observed data is not needed. This is used for ungauged basins, often parameter values from similar basins are used for calibration by an expert who has knowledge of a neighbouring catchment with similar characteristics. Catchment modeling must take care of parameters like rainfall, losses in the catchment. While assessing rainfall, the temporal and spatial patterns must be considered. Whether the storm is sub areal in nature, or if it occurs in a large region then areal reduction factors also need to be considered. Difference in storm characteristics alters the catchment's response to the storm; any model must be able to capture this aspect. In case of catchment rainfall runoff model, it is important to consider antecedent moisture conditions in the catchment in the 5 days prior to the storm event. For example, the storm event considered may be of 24-hour duration, the frequency may be yearly, 2 yearly, 5 yearly, 10 or 50 years return period for that matter. Thus, the time stepping in modeling is important to fix for any type of model.

Continuous Modeling

In this case a continuous time series of rainfall events and other climatic factors are studied in order to generate complete stream flow time series using the catchment rainfall runoff response. The continuous modeling brings out the variability in flow pattern, for that variability of input patterns must be available or can be simulated. Innovative storm water management techniques need to be based on continuous models so that the constraints in implementation may be understood and only then implementation shall be successful.

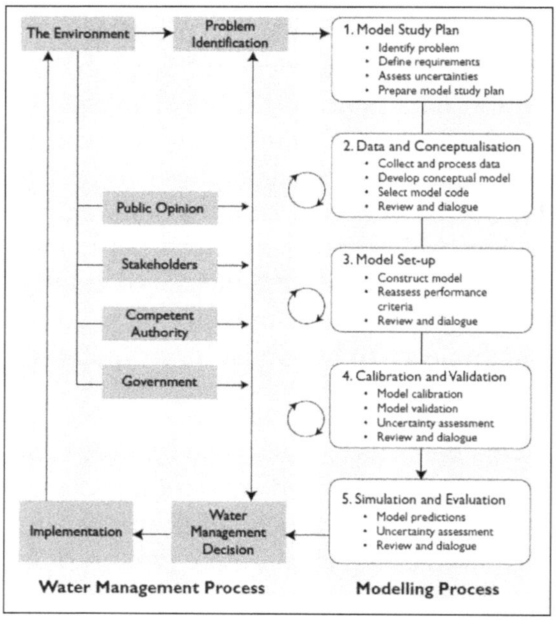

Model Development.

The process of model development is best understood from figure, the development involves five important steps as shown in the figure. The environment may be taken to mean the entire catchment characteristics broadly. Once the problem is identified a model study plan is developed, then the data collection plan is prepared and conceptualization takes place. After this a model is constructed and then the steps of calibration and validation follow. As seen from the figure the steps of data and conceptualization, model set up and calibration and validation are iterative in nature. For any model to be successful it has been identified that stakeholders are involved, competent authorities are involved and public opinion is considered. All this strengthens the model and makes it robust. Once calibration and validation of the model is done then simulation takes place, this helps to check the efficacy of the model. It is only when the users of the model can take suitable water and eco system management decisions based on the model output, that the model can be considered successful.

Model Purpose

There are a numbers of reasons and purposes for which a model needs to be developed. So it is important to identify the purpose for which the model is to be developed. For example, the rainfall runoff relationship is studied in catchments extensively; however, the purpose for which it is being studied may be different depending on the problem that we are trying to solve. The purpose for which the relationship is to be studied could be for any of the following purposes in a catchment (may not be limited to these only):

- To build rain water harvesting structures (ponds, check dams and the like) so that regular floods or droughts occurring in the particular catchment can be addressed, so it is important to quantify the surface runoff rate. Based on this and the time period, the volume generated will determine the size of the structure.
- To allocate water in the catchment based on the various water demands, like agricultural water demand, industrial and urban water demand, so it important to understand how much water can be extracted from surface water sources and ground water sources. It is also used to decide when surface water may be used and when ground water may be used. This strengthens the cause of conjunctive use.
- Forecasting of events when flooding/waterlogging may occur in the catchment, when a certain threshold of precipitation is crossed, the runoff generated causes flooding in the catchment, erosion that may occur because of severe rainfall, quantity of sediment generated.
- All the three purpose categories mentioned above are part of the water management process in the catchment.

Objectives of the Model

Once the model purpose is defined, the objectives of the model will need to be specified. It is imperative to see who are the people or stakeholders who will be using the model, how will the results of the model be used (application). In order to get a clear perspective of these, the specific outputs needed from the model should be discussed with the users. For example, if the objective of the model is to arrest stream bank erosion problem within the catchment then the model must be able to suggest vegetation measures to be planted on the banks, management measures needed to sustain them and other initiatives that may be taken. For this the annual sediment load, quantity of erosion taking place, nature of rainfall, runoff generated and soil characteristics and vegetation characteristics within the catchment must also be known. Prior knowledge of upstream catchments will also be helpful if the stream is coming from outside the catchment.

Data Collection

Once the purpose and objectives are set, what data already exists must be explored. For this a study plan is prepared and pre- existing information is explored. To go back to the rainfall runoff model, measured data on stream flow if available may be collected (how many years stream flow data is available that are to be seen, if rainfall records are available, that too should be collected). Next the quality, accuracy and completeness of data are to be evaluated. Major constraint faced during the data collection stage is the fact that the model may be designed for different time steps while the spatial and temporal data are available in different temporal and spatial resolution. Analysis has also to be done to determine how to make the available data useful to the model scale, for example how to carry out downscaling of temperature and rainfall data from global or regional climate models (which are used for prediction of future temperature and precipitation trends).

One has to analyse if sufficient expertise is available to analyze and assess such data and how much work is involved. If the data cannot be easily collected then may be the objectives of the model may be revisited. Other kinds of data sets that are used for analysis in catchment modeling are land use data, probability distributions of precipitation, nutrient load, sediment load. In addition data set which is of importance is the knowledge about natural processes in the catchment like direction of stream flow, nature of stream flow during different seasons. To take the example of rainfall runoff model for a catchment, in order to build the model well, if the rainfall runoff data record if available on a daily basis then this gives an idea about storage characteristics of the catchment. In summers and at the onset of monsoons the evapotranspiration, soil moisture, infiltration are all quite high, while during the middle of the monsoon months, the effective rainfall which causes runoff is much higher as the storage facilities are full to the brim. This analysis of effective rainfall and therefore surface runoff helps to build a more robust model.

Conceptualization

At this stage it is important to identify the types of variables to be covered and the expected output based on the data availability, so the decision regarding whether the model will be empirical, conceptual, physical, lumped or distributed, event based or continuous, probabilistic, stochastic or deterministic can to be taken. Once the conceptual model is chosen another set of iteration with the objectives, plan, data availability is carried out involving the various stakeholders' needs to be carried out so that all the hypotheses may be checked.

After conceptualization of the entire system (in this case the catchment), design for a site specific model begins. The model at this stage is represented in mathematical terms. So variables have to be defined, a variable is one which changes with time (in our case the rainfall event measured in mm or cm changes with time). Variables normally are measured in units of mass, volume, length and time. The other components of a model are constants, parameters and mathematical operators. Constants have fixed value in the model which does not change with time while parameter values may vary with time, they are either measured or are inferred from secondary data (for instance, hydraulic conductivity used for groundwater models are either measured through experiments or are obtained from secondary data depending on properties of the aquifer). Mathematical operators like addition, multiplication, division, subtraction express the relationship amongst variables.

The chosen model structure is guided by prior knowledge of science in the relationship between the variables for which the model is being developed and the governing principle is to keep the model simple. If it is possible to explain the relationship with one dimensional equation, two and three dimensional are very rarely used, one dimensional differential equation will be used for example instead of three dimensional differential equation if there is scope to explain the relationship adequately with this. Every model has a few assumptions which are justified based on the understanding of the natural system of the area under study (for example, usually point rainfall data is available this is used to derive catchment area average rainfall data based on certain assumptions or assigned weights that are given to different point rainfall station data).

Model Performance

Once the model is set up (mathematical equation is identified), variables, parameters and constants are identified and assigned values, the model is run and the performance of the model is evaluated. The model performance is evaluated essentially based on the variation in measured values and model simulated values. Often the model cannot predict the extremes of highs and lows with equal precision. For example in case of rainfall runoff model when the model results of stream flow is compared with observed stream flow results there will be variation in the low flows and high flow events. This results in large residual differences between predicted (model values) and observed stream flows. This is because often the measured low stream flow data at stream gauging stations are unreliable; same may be the case with recording high stream flows during flood events. Typically the difference between observed and modeled stream flows for a stream flow at a catchment gauging station is shown below in figure, as shown both for high and low flows the difference is high.

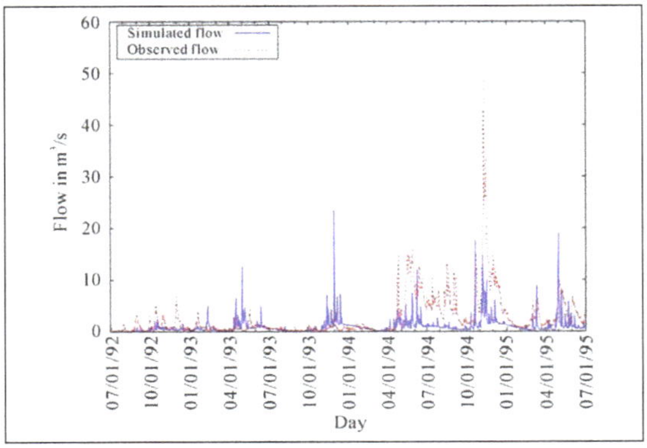

Simulated flow through model and observed flow.

Measurement of Performance of a Model

The root mean square (RMS) is a measure of error or difference between observed and modeled values. It is calculated as shown below:

$$\text{RMSE} = \sqrt{\frac{\sum_{i=1}^{n}\left(X_{obs,i} - X_{model,i}\right)^2}{n}}$$

where X_{obs} is observed values and X_{model} is modeled values at time/place i.

The errors have the same units as the data measurements, therefore if it is stream flow the unit of measurement will be m³/day or m³/sec as the case may be. This method works best for high value deviations because of the squaring effect, than the lower values. The Nash and Sutcliffe efficiency (NSE) is also a measure of the performance of hydrological models.

$$E = 1 - \frac{\sum_{t=1}^{T}\left(Q_o^t - Q_m^t\right)^2}{\sum_{t=1}^{T}\left(Q_o^t - \overline{Q}_o\right)^2}$$

Q_o, Q_m are observed/measured discharge and model discharge values. $\overline{Q}_o$ is mean observed discharge for the time period T over which the model values are tested. Nash-Sutcliffe efficiencies can range from $-\infty$ to 1, an efficiency of E=1, means that the observed and model values have a perfect match, while E=0 indicates that the models predictions are similar to the mean of the observed data. However, if the value of E is negative ($-\infty < E < 0$), it means that the observed mean is a better predictor than the model values of discharge. It means the model is unable to simulate the stream flow and iterative steps must be repeated to revise the model equation. In short closer the value of efficiency is to 1, more accurate is the model. WHAT-Web based hydrograph analysis tool may be used to carry out model performance.

Once the model code logic is verified and model behavior is accepted a sensitivity analysis is carried out. If the sensitivity analysis reveals over parameterization, which is indicated if the sensitive analysis reveals that the model is insensitive to a few parameters, then it is better to remove those parameters and simplify the model. Saltelli et. al. give an idea about various methods of sensitivity analysis and their necessity. At this stage review and dialogue with relevant stake holders may be carried out and if need be further iterations can be carried out. Once the performance of the model is established, then the step of validation is taken up.

One of the model validation methods is explained here. Models are widely used to understand ungauged catchments as the catchment data is insufficient. So what is done is two similar catchments are chosen (say X and Y), which have similar properties between them and similar to the ungauged catchment. The model is calibrated using data (e.g. stream flow data) from one and validated using the catchment data of Y. Only if the two validation results are acceptable, then the credibility of the model is established to simulate stream flow in the ungauged catchment too. However, it is not always possible to have sets of data for validation of model especially for ground water contaminant problems, as there may not be any previous history of contamination in the catchment. That does not mean that the model is not acceptable, only that a greater degree of uncertainty has to be built in the model output, as the historical dataset is not available or is available for a very short period in time.

Under such circumstances, a matrix is used to understand the sources of uncertainty in the modeling process, it provides an identification of all relevant sources of uncertainty and a prioritising based on a qualitative assessment of their importance. The matrix provides a framework to keep track of all sources of uncertainty during the modeling process. The source of uncertainty could be in the input data, model structure, parameters chosen, numerical method or software chosen. In order to understand the success of the model, following points must be satisfied:

- The purpose of modeling is satisfied.
- The model behavior is approximately the same as the catchment system under study.
- The space and time steps chosen in the model are sufficient to understand the real catchment scale problems.
- The model boundaries are chosen correctly.
- The model implementation follows all steps including calibration and validation if possible.
- Sensitive parameters must be identified.
- Uncertainty analysis must be performed.

Role of Models in Decision Making Process

Scientists now have a more clear understanding of the physical complexities of a catchment and river basin, however, it is not enough to understand the catchment only scientifically today. The level of public involvement is high and the objectives of the models have to be communicated to the stake holders as well. Decision support models play a significant role in that. The scientific part of the model helps in assessing the evapotranspiration requirements of the crop through analysis of climate data, and then the other interface of the model looks at optimum use of water (water allocation) through irrigation scheduling. Should less water be available in a particular season due to drought, it allows the model to decide deficit irrigation techniques. All stake holders can be communicated the output of the model and a collective decision taken. It is one of the most successful decision support methods that are used now to choose a sustainable cropping pattern in different agro climatic zones of the world. The key issues related to use of quantitative models such as CROPWAT as a decision support method may be as follows:

- Before adopting quantitative methods, qualitative thinking has to take place. Only then, the decision making process can be made robust with the help of the model.
- The quantitative model may not be able to take care of all the stakeholders in the catchment. However, all stakeholder groups should be involved in the decision-making process as far as possible.
- At all times, it must be remembered that a model is a simplified version of reality. No model can perfectly represent the complex natural catchment system completely. It is thus important that the assumptions must be clearly stated in the model.
- Models today are sophisticated with high computational power, however, quantitative models can do only what they are specified to do. Therefore, it is important that the objectives of the model are clearly specified. Only then the model outputs can be clearly used in the decision-making process. For example, the rational formula gives peak discharge which the probability of exceedance or return period that is mentioned (flood return period of 50 years,100 years) which is then used to design structures for flood protection or proper channelization so that minimal waterlogging takes place.

River Basin Modeling

A river basin system is made up of three basic components: water source components, in-stream and off-stream demand components, and intermediate components. Water source components are rivers, canals, reservoirs, and aquifers; off-stream demand components are irrigation fields, industrial plants, and cities; in-stream demand components are hydropower, recreation, environment; and intermediate components are treatment plants, water reuse and recycling facilities. Water management involves development, control, protection, regulation, and beneficial use of water sources by meeting up the water demands in an efficient way. With increasing competition for water use across sectors and regions, a river basin has been recognized as the appropriate unit of analysis for addressing the challenges of water resources management. The essential relations within each component and the interrelations between these components in the basin can be considered in an integrated modeling framework which can provide essential information for policymakers in their resource allocation decisions.

Simulation and optimization are the two principal approaches to river basin modeling. In simulation water resources behavior are simulated based on a set of rules governing water allocations and infrastructure operation, and in optimization, the allocations are optimized based on an objective function and accompanying constraints. Water resources management involves the identification and development of water resources project investments that are net benefit maximizing or at least cost-minimizing, while considering non monetized impacts, such as potential ecosystem degradation or negative social impacts.

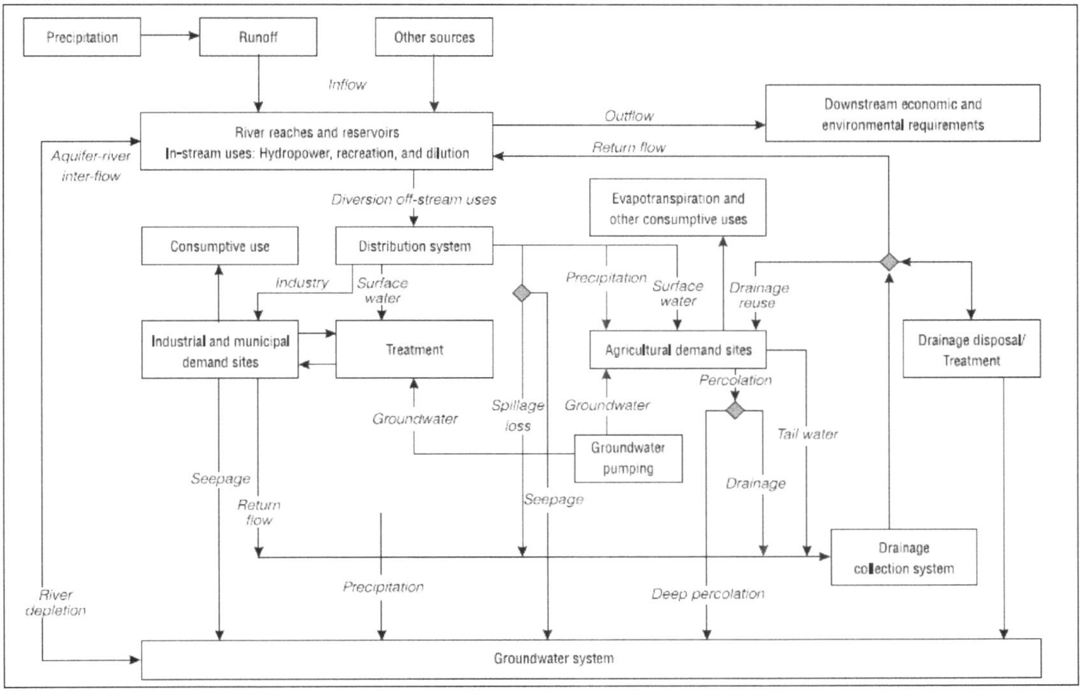

Schematic representation of river basin process showing its components: the water supply system (groundwater and surface water), the delivery system (canal network), the water users system (agricultural, municipal, and industrial), and the drainage collection system (surface and subsurface).

Researchers reviewed modeling of water resources management at the basin level, Mayer and Muñoz-Hernandez reviewed integrated water resources optimization models, Hajkowicz and Collins reviewed multiple criteria analysis for water resource planning and management, Rani and Moreira reviewed various optimization techniques used in water resources management modeling for reservoir system operation.

Maximizing Economic Benefit Approach

In this approach, the river basin model is developed as a node-link network representing the spatial relationships between various physical entities in a river basin where nodes represent river reaches, reservoirs, and demand sites, and links represent the connections between these entities. Generally, the modeling framework includes three components: hydrologic components (e.g. reservoirs, river reaches, crop fields etc.); economic components (e.g. calculation of benefits from water uses); and institutional rules and economic incentives that affect the hydrologic and economic components.

The general objective of this type of management model is to allocate water based on maximizing the annual net profits from crop or agricultural water use, water used in aquaculture production, residential water use, industrial water use, hydroelectric power use, water allocated for ecosystem functioning, and recreational use etc.

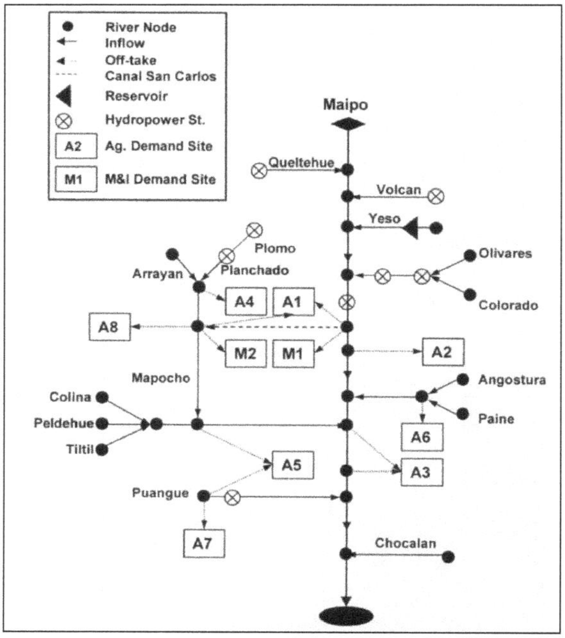

Node-link network of the Maipo River Basin in Chile, representing the spatial relationships between the physical entities in the river basin.

The general equation of finding the optimal allocation can be,

$$\max_{X \in \phi_X} f(X)$$

where, f is the objective function, X(s, Q) consists of state variable vector s and decision variable vector Q and $\phi_X = \phi_s \cup \phi_Q$ is the feasible region of X represented by a set of constraint equations, ϕ_s is the feasible region of s and ϕ_Q is the feasible region of Q. The decision variables typically include flows associated with the water use or allocation categories of interest and can be expressed as,

$$Q = \{Q_C, Q_A, Q_R, Q_H, Q_E, Q_{REC}\}$$

where Q is the vector of water withdrawals and $Q_C, Q_A, Q_R, Q_H, Q_E, Q_{REC}$ are crop or agricultural water use, water used in aquaculture production, residential water use, industrial water use, hydroelectric power use, water allocated for ecosystem functioning, and recreational use, respectively.

Considering maximization of economic benefit, the optimization framework represented in above equation can be expressed as,

$$\max_{X \in \phi_X} EB = \sum_i EB_i(Q_i, s)$$

where the economic value EB_i is associated with water withdrawal Q_i from water use sector i.

Equation above is a general expression to maximize the economic benefit of a water resources system. Examples of finding the economic benefit from different sources of water use have been applied in various water resources management studies are listed below:

1. Economic benefit from agricultural water use:

$$EB_c = \sum_{i_c} Y_{i_c}\left(Q_{i_c}\right)\left[P_{i_c} - C_{i_c}\left(Q_{i_c}\right)\right]$$

where $Y_{i_c}, P_{i_c}, Q_{i_c}$ are the yield, selling price and production cost of crop i_c respectively.

2. Economic benefit from aquaculture production:

$$EB_A = \left[F_i \cdot a \cdot \left(P_i - C_i\right)\right] \cdot F1_i \cdot F2_i$$

where F_i, P_i, C_i are the fish production, fish price and fish production cost at a demand site i; a is a parameter relating normal to estimated maximum fish production; $F1_i$ and $F2_i$ are the lowest monthly factors relating actual and maximum instream flows and actual and maximum lake storage.

3. Economic benefit from residential water use:

$$EB_R = \int_{Q_R}^{Q_{R,O}} P_R dQ_R$$

where Q_R, $Q_{R,o}$, and P_R are the water demand, initial water demand and water price for residential water uses respectively. The relationship between water demand and price is often given by,

$$Q_R = cP_R^{\varepsilon}$$

where ε is the elasticity and c is a constant.

4. Economic benefit from industrial water use:

$$EB_I = \int_{Q_I}^{Q_{I,O}} P_I dQ_I$$

where Q_I, $Q_{I,o}$, and P_I are the water demand, initial water demand and water price for industrial water uses respectively. The relationship between water demand and price can be given as,

$$Q_I = Q_I\left(P_I, P_Y, Y\right)$$

where, Y and P_Y are the amount of product generated and cost associated with all other cost respectively.

5. Economic benefit from hydropower production:

$$EB_H = Q_H H_e E_f (P_H - C_H)$$

where Q_H, H_e, E_f, P_H, C_H are the released flow, effective head, generator efficiency, selling price of generated electricity and cost of generated electricity respectively.

6. Economic benefit associated with environment:

$$EB_E = \sum\nolimits_t BE.WECO^t$$

where BE is the socioeconomic net benefit per unit of ecological water use and $WECO^t$ is the amount of water for ecological use at a period of t.

7. Economic benefit associated with recreation:

$$EB_{REC} = \sum\nolimits_i \left(b_i^o + b_i^1 Z_{it} + b_i^2 Z_{it}^2\right)$$

where Z_{it} is the storage of ith reservoir at period t and b is a constant.

Researchers developed a basin scale integrated economic-hydrologic water modeling system to account for the interactions between water allocation, farmer input choice, agricultural productivity, non-agricultural water demand, and resource degradation to estimate the social and economic gains from improvement in the allocation and efficiency of water use in Maipo river basin. Their modeling framework includes hydrologic components, water use components and economic components. The objective was to maximize economic profit from water supply for irrigation, municipal and industrial water use, and hydroelectric power generation, subject to institutional, physical, and other constraints.

Researchers presented an Interyear Control Program (IYCP) and a sequence of yearly models (YMs) as a framework for sustainability analysis in water resources management and application to the Syr Darya Basin of central Asia. YM is a short-term optimization model that maximizes total water use net benefit in a river basin for a year. It is based on a node-link network, with source nodes such as rivers, reservoirs, and groundwater aquifers and demand site nodes such as agricultural, municipal, industrial, and ecological demand sites and hydropower stations. Objective function of these yearly models was to maximize the total net benefit of water use in any year from irrigation, hydropower and ecological flows. Irrigation benefit was calculated by subtracting the withdrawal water price and associated with the salt discharge tax from the irrigation profit. They assumed the hydropower generation as a linear relation with water release through the turbines, and the hydropower profit was estimated by calculating net benefit from unit release of withdrawal water. The economic benefit of ecological water use was estimated from the amount of water for ecological use and the socioeconomic net benefit per unit of ecological water use.

Their long term model (IYCP) used sustainability criteria to control relations between short-term irrigation practices and their long-term socioeconomic and environmental consequences. The long-term objective function was a linear combination of sustainability criteria which include risks associated with agricultural and ecological water supply, environmental integrity, temporal and spatial equity, and economic acceptability. They optimized the ratio of the marginal improvement in economic benefit resulting from increases in water use efficiency to the corresponding marginal increase in infrastructure investment needed to produce the given level of water efficiency increase along with the objective of maximizing overall economic benefit.

Researchers applied the similar approach to develop the short term integrated hydrologic-agronomic-economic model for Syr Darya river basin management considering both the surface water and ground water as the source. Researchers analyzed the relationship between physical and economic efficiency of irrigation water management at the basin level based on an optimization model developed for the Maipo River Basin in Chile. The optimization model was used to maximize the total net benefits to water use (TNB) from in-stream environmental water uses, hydropower generation, irrigation and municipal and industrial water uses. The formulation of net profit includes profits from irrigation, municipal, industries, and hydropower; cost of withdrawal water for these sectors; the return drainage flows and costs for drainage treatment from the irrigation, municipal and industrial sectors. They did not explicitly valued environmental uses in their modeling framework; however, a minimum flow requirement for environmental purposes was incorporated as a constraint.

They set three different scenarios in analysis: fixed irrigation technology scenario which explores the outcome for physical and economic irrigation efficiency of alternative economic incentives; fixed water rights scenario which examines changes in physical and economic efficiency at the basin scale under different levels of irrigation system efficiencies; tradable water right scenario which shows the impact of different levels of water withdrawals on physical and economic efficiency with a given water requirement. They compared the results with the full optimization scenario which eventually assumed omniscient decision-maker allocating water to the most profitable uses and found that the optimization scenario generates the highest total benefits from water usage in the river basin.

Researchers developed long term and short term decision model for integrated assessment on water allocation options in the Namoi River catchment, Australia. The decision model consists of three components: the policy component, agricultural production extraction economic component and the hydrologic component. They mapped the entire basin into a number of relatively homogenous regions considering groundwater policy, surface water policy and, production type. Their short term decision model is used to maximize the profit from the agricultural production as,

$$\text{Max}\,\pi_{\alpha,t} = \sum_{i=1}^{i=I} \sum_{j=1}^{i=J} \left(P_{ij} y_{ij} - c_{ij}\right) a_i p_{ij}$$

where $\pi_{\alpha,t}$ is the short term profit from agricultural production for region α at time t; i,j,I,J are the crop activity, product type, total crop activity, total product type; P_{ij}, p_{ij}, y_{ij}, c_{ij}, are the price, crop rotation proportion, yield and cost of production of product j from crop activity i respectively; a_i is the area of land devoted to crop activity i. The long term decision model was used to maximize an objective function based on the short term profit, interest rate, capital cost, and irrigation technology option.

Researchers calibrated a holistic water resources–economic model involving essential water resources and economic components in a consistent model for Maipo River Basin, Chile. They evaluated economic benefits from water uses for different demand management instruments (e.g. markets in tradable water rights, production and benefit functions with respect to water for the agricultural and urban-industrial sectors etc). The optimization model was used to maximize water use profit with embedded basin-wide simulation of flow and salinity balance, reservoir operation, irrigation scheduling, and crop growth. They calibrated the model to match a large number of computed values with observed or recorded values under given climate regime and agricultural inputs (e.g. flow through river reaches and reservoirs; water withdrawals for both agricultural, municipal and industrial sites; crop harvested area and yield; farmer incomes and profit from industrial and municipal water uses etc).

Introduced the role of wetlands and fisheries in the economic hydrologic management of Mekong River Basin. They described valuation techniques for wetlands and fisheries in a basin context and estimated the economic benefits associated with aquaculture production by a production function approach, where the value of the fisheries is related to instream flows. River's hydrologic water balance was used to estimate the water supply, while water demand is estimated endogenously based on functional relationships between water and productive uses in different sectors (e.g. irrigated agriculture, domestic-industrial areas, wetlands, fisheries, and hydropower). Optimization was based on maximizing economic benefits of water use while keeping the balance between water supply and demand.

Researchers used a holistic approach in analyzing the water allocation to generate options for water policy reform in the Dong Nai River Basin for optimal, sustainable, efficient and equitable utilization of water resources. There were three components in the modeling framework: hydrologic components which include the water balance in reservoirs, river reaches, and crop fields; economic components which include the calculation of benefits from water uses by sector, province, and demand site; and institutional rules and economic incentives that affect the hydrologic and economic components. As a similar approach to Ringler and Cai water supply is determined through the hydrologic water balance in the river system, and water demand is determined endogenously within the model based on functional relationships between water and productive uses. The optimization objective was to maximize economic benefits to water uses in irrigated agriculture, domestic and industrial uses, and hydropower and environmental requirements to control saltwater intrusion were included as flow constraints.

Researchers developed a spatially distributed simulation-optimization model to investigate conjunctive management of surface water and groundwater resources to alleviate the effects of droughts on the profitability and sustainability of irrigated agriculture in Yaqui Valley, Mexico. The simulation model consists of an agronomic component and simulators describing the hydrologic system and profit maximization problem was formulated and solved using large-scale constrained gradient-based optimization. The decision variables of their objective functions include crop area, groundwater pumping and reservoir release. Optimization was based on maximizing the profit from irrigation district while keeping the reservoir spills minimum.

Researchers designed and constructed an integrated basin wide nonlinear programming model for the purpose of optimizing water allocations in Rio Grande Basin. The objective of their study was to conduct an integrated economic-hydrologic-institutional analysis of policy responses to mitigate drought impacts. Their integrated model was formulated as a mathematical optimization problem, using the sum of benefits from basin water diversions for off-stream uses and benefits of in-stream use recreation as the objective. Total water-related benefits were estimated as a quadratic function of consumptive use and recreational benefits of reservoirs were assumed as a quadratic function of storage.

Researchers presented a modeling framework for the economic optimization of conjunctive use in a water resource system by efficiently integrating distributed parameter groundwater simulation and dynamic stream-aquifer interaction within the overall model for the Adra River system in Spain. They defined the objective function to be minimized as the sum of the economic losses derived from water shortages in the consumptive demands plus the variable operating costs such as pumping costs in order to optimize the net benefit of water use in the system.

Table: List of selected water resources management studies based on maximizing economic benefit.

Application Area	Maximizing Net Benefit Components	Optimization Technique
Maipo River Basin, Chile	IRG, HYP, MUN, IND	NLP
Syr Darya Basin, Central Asia	IRG, HYP	GA, LP, DP
Maipo River Basin, Chile	IRG, HYP, MUN, IND, ENV	DP
Namoi River catchment, Australia	IRG	LP, DP
Maipo River Basin, Chile	IRG, HYP, MUN, IND	GA
Mekong River Basin, Southeast Asia	IRG, HYP, MUN, IND, ENV, FIS	NLP
Dong Nai River Basin, Vietnam	IRG, HYP, MUN, IND	NLP
Yaqui Valley, Mexico	IRG	NLP

Rio Grande Basin, USA	IRG, MUN, IND, REC	NLP
Adra River Basin, Spain	IRG, MUN, IND	NLP
Abbreviations for management component: IRG=Irrigation, HYP=Hydropower, ENV=Environmental, MUN=Municipal, IND=Industrial, FIS=Fisheries, REC=Recreation; abbreviation for optimization techniques: GA= Genetic Algorithm, LP=Linear Programming, DP=Dynamic Programming, NLP=Non Linear Programming.		

Minimizing Cost Flow Approach

In this approach, the modeled region is translated into a network composed of nodes interconnected by flow links to represent physical channels. Each node (other than junction) or flow link in the network represents one of the model components. Nodes are used to represent points of inflow, diversion, storage or consumption. Flow links define the direction of surface water movement from one location to another (e.g. river reaches, diversion canals, and return flow channels).

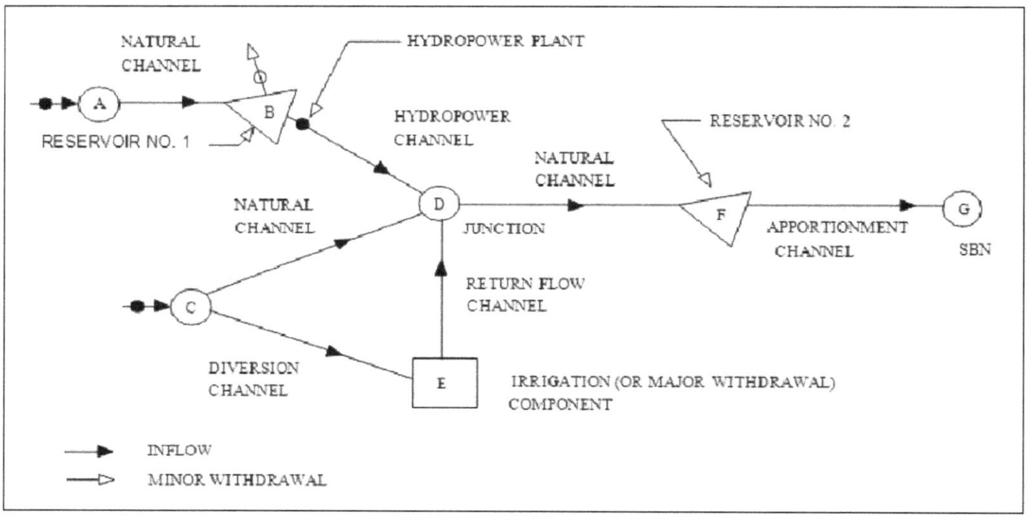

Schematic diagram showing node link system.

Water licensing system or priority of supply is set to various water user groups and water resources allocation is made to maximize supply to all users according to their respective priorities, hence maximizing the economic value of agricultural and urban water use. All flows in the network are weighted by a unit cost that can vary between arcs and optimization is done by minimizing the overall cost of flow. Costs include convex economic losses to agricultural and urban regions, urban water quality costs for salinity and treatment, and pumping and other operating costs.

Considering a network consisting of ordered pairs (i, j) of arcs A and a total of N nodes i, j. The minimum cost flow problem is defined as,

$$\min \sum_{(i,j)\in A} c_{ij} x_{ij} \quad \forall i, j \in N$$

Subject to: $\sum_i x_{ij} - \sum_i x_{ij} = 0 \forall \ i \in N$

$$0 << l_{ij} << x_{ij} << u_{ij} \ \forall(i,j) \in N$$

where, where c_{ij}, l_{ij}, x_{ij}, and u_{ij} are respective cost factors per unit of flow, lower bound, flow, and the upper bound on flow along an arc (i, j).

Labadie developed a generalized river basin network model, MODSIM, which employs a state-of-the-art network optimization algorithm for simultaneously assuring that water is allocated according to physical, hydrological, and institutional aspects of river basin management. Within the confines of mass balance throughout the network, MODSIM sequentially solves linear optimization problem over planning period of record through an efficient minimum cost network flow program. Application of MODSIM includes Rio Grande River Basin; the Poudre River Basin in Colorado the Upper Colorado River Basin the Upper Snake River Basin and the Guiem River Basin in South Korea as KMODSIM.

Table: List of selected water resources management studies based on minimizing cost flow.

Model Name	Application Area	Optimization Technique
CALVIN	California, USA	NFP
CALSIM, WRIMS	California , USA	LP
WRMM	South Saskatchewan River Basin, Western Canada	NFP
MODSIM	Guiem River Basin in South Korea	NFP
MODSIM-PSO	Sirvan basin, Iran	SO
Abbreviation for optimization techniques: LP=Linear Programming, Stochastic Optimization=SO, NFP=Network Flow Programming.		

Alberta Environment developed the Water Resources Management Model (WRMM) as a planning tool for water resources utilization, with a river basin being the fundamental unit of study. WRMM is a deterministic, steady state, surface water allocation model which is a multipurpose, multi reservoir simulation model with a nested optimization subprogram. Water allocation priorities are defined by a penalty point system and optimize technique is used to minimize the overall system penalty.

WRMM enables easy and repeated analysis of the response of the river basin to differing natural conditions and planning alternatives. The model can represent different physical processes as water supply from storage, headwaters, local runoff, reservoir storage, release, precipitation and evaporation, natural streams, irrigation consumption, hydropower production; major and minor consumption etc. The principal inputs of WRMM are the river natural flow data, irrigation consumptive use, non-irrigation withdrawals, instream objectives, reservoir and canal structure capacities, license priorities etc. Principal model outputs are: time series of reservoir levels, regulated channel flows, and irrigation or industrial consumption.

Each WRMM run can be considered as an operational policy or structural alternative and the simulation outputs are then compared for various components in order to assess the impact of alternative policies or design proposals. Each river is split into a set of reaches along which flow is kept constant in the model, and the runoff contribution along the reach is represented as the local inflow at the downstream end of the reach. A major withdrawal can represent industrial or municipal consumptive use that does not always have to be met, depending on the penalty assigned to it and the availability of water. Return flow channels were incorporated in the WRMM to allow modeling the return flow diverted flow back into the main watercourse from irrigation and industrial use.

WRMM has been using a planning tool in management of surface water resources in South Saskatchewan River Basin of Western Canada. Researchers presented an economic-engineering optimization model, namely CALVIN, which comprises of California's major water supply system and demonstrates the feasibility of using economic-engineering optimization for the planning of California's water resources. CALVIN model is used to optimize water facility operations and allocations that maximize the economic value of agricultural and urban water use. Economic value functions for year 2020 levels of development was used to represent agricultural and urban water demands and monthly operation and allocation decisions are made for a 72-year period based on the 1922–1993 hydrologic period. Environmental flow requirements as well as facility capacities and flood control operations are used as constraints. Surface water, groundwater, and return flows are considered as the source of water.

A separate optimization model, SWAP (Statewide Agricultural Production model) was used to derive economic benefit and loss functions for farm water use. SWAP maximizes farm profit for each agricultural demand area, given a quadratic crop production function with water, land, technology, and capital inputs, and constraints on water and land availability. Economic demand curves for urban water use are used to estimate economic losses from urban residential water scarcity; industrial water scarcity costs are taken from a statewide study; and commercial and institutional water demands are assumed fixed. CALVIN is also used in analyzing long-term water policy, planning, and management in California.

Researchers applied California Water Resources Simulation Model (CalSim, currently known as Water Resources Integrated Modeling System or WRIMS), a general purpose reservoir-river basin simulation model in California State Water Project (SWP) and the federal Central Valley Project (CVP). In CalSim, user specifies a series of objectives in the form of relative priorities for water allocation and storage while physical capacities and specific regulatory and contractual requirements are input as linear constraints on system operation. For each time period, the objective function is maximized to determine a solution that delivers or stores water according to the specified priorities and satisfies system constraints.

Researchers presented a methodology for optimized design and operation of the upstream Sirvan basin in Iran by integrating MODSIM generalized river basin network flow model and Particle Swarm Optimization (PSO) algorithm. The design variables (e.g. the size of planned dams and water transfer systems), operational variables (e.g. the relative priorities for meeting reservoir target storages) are varied and evolved using PSO algorithm in the developed PSO-MODSIM model. The system performance are simulated and the fitness of each set of those design and operational variables with respect to the model's objective function are evaluated by MODSIM and PSO objective function is used to maximize the total net benefit due to supplying water to different types of water uses considering construction costs of dams and water transfer and/or pumping systems. The PSO-MODSIM model is used to size the planed water storage and transfer components of a river basin system and to allocate water resources optimally over time and space among competing demands, considering coordinated operation of the system components.

System Dynamic Modeling Approach

In System dynamics (SD) modeling approach qualitative and quantitative causal models are created to capture the interrelationships of the physical (e.g. water inflows, outflows) and behavioral (e.g. decision rules, perceptions) processes in the system. It is a rigorous methodology for thinking, visualizing and communicating about the future evolution of complex systems which provides a feedback-oriented modeling framework for learning and communicating about the inherent complexity of water management and has been widely applied in many environmental problems, including water management. Basic equations showing the mathematical form of system dynamics are as follows:

- The evolution of system is given by,

  $$\text{Levels}_t = \int_o^T \text{Rate}\,dt$$

- Computational rates determining the evolution is given by,

  $$\text{Rates}_t = f(\text{Levels}_t, \text{Aux}_t, \text{Data}_t, \text{Const})$$

- Intermediate results necessary to compute rates is given by,

  $$\text{Aux}_t = g(\text{Levels}_t, \text{Aux}_t, \text{Data}_t, \text{Const})$$

- Initialization of the system is given by,

  $$\text{Data}_t = f(\text{Levels}_t, \text{Aux}_t, \text{Data}_t, \text{Const})$$

Where, 'Levels' used to represent accumulation, 'Rates' used to present change per unit time of state variable, auxiliary variables ('Aux') used to store and constants ('Const') do not change with time. f, g and h are arbitrary, non-linear, potentially time varying and vector-valued functions.

SD modeling consists of different phases as, problem definition, system conceptualization, model formulation, model evaluation/testing, policy analysis and implementation. In problem definition stage, the main issues that are relevant to water resource management are identified and a conceptual framework of these issues categorized into supply, demand and price adjustment is generated. Based on the conceptual framework, a causal loop diagram (CLD) is used to capture the interactions between supplies, demands and economic development and the feedback structure is developed to represent the influence factors.

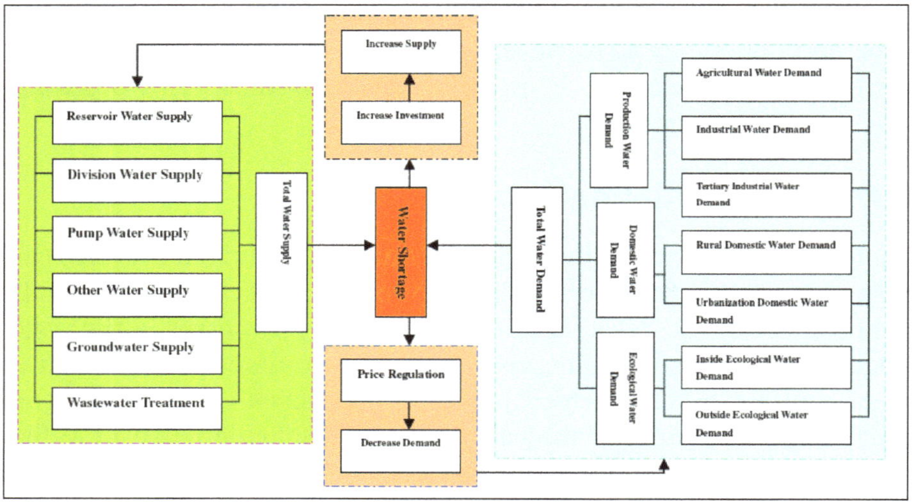

(a) Conceptual diagram

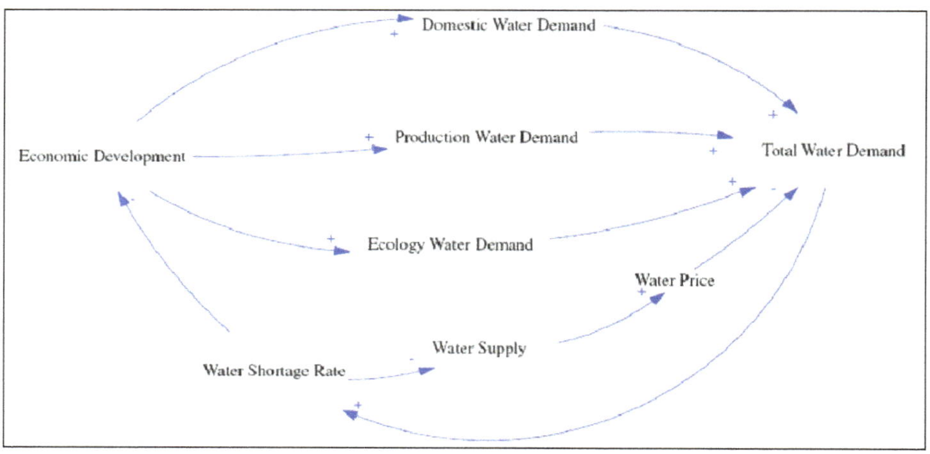

(b) Casual loop diagram for water resources management by System Dynamics Modeling Approach.

Influences is considered as positive if an initial change is amplified in the same direction or negative if an increase in one element causes a decrease in another. Figure shows the conceptual diagram identifying various supplies, demands and economic interactions; and Figure b shows corresponding casual loop diagram for water resources management in Yulin city, China by system dynamics modeling approach. In model formulation stage the key elements are quantified as variables and their influences are formulated mathematically. A wide number of water resources management studies based on system dynamics modeling are found in literature.

Researchers developed Water Resources System Dynamics (WRSD) model to evaluate the sustainability of the water resource system and to simulate the system by capturing the dynamic character of the main elements affecting water demand and supply in the Yellow River basin. The total demand is calculated by summing the following demands: industrial water demand in city, town and rural area; domestic water demand in city, town and rural area; the agricultural water demand; the environmental and ecological water demand; and the inter-basin water transfer. Total available water-supply is calculated from available surface water supply, available groundwater supply and re-used treated waste water-supply. Sustainability index, which is the ratio of aggregated possible water deficit relative to the corresponding supply in the same region, was used to evaluate the performance of water resource systems.

Table: List of selected studies of water resources management based on system dynamics modeling approach.

Application Area	Application Purpose
Yellow River basin, China.	Evaluation: The sustainability of the water resource system.
Middle Rio Grande river basin, USA.	Community-based water resources planning project.
Red River basin in Manitoba, Canada.	Flood management.
Hypothetical	Improving the efficiency of water allocation in irrigation system.
Tianjin, China	Policy-making on water resources allocation and management.
Yulin City, China	Effectiveness of various supply and demand policies in meeting socio-economic and ecological requirements.

Researchers applied system dynamics modeling to describe a community-based planning project in the Middle Rio Grande river basin which served as a catalyst for the preparation of a 50-year water plan including the definition and communication of water resources issues, the education of the public about the complexities surrounding the regional water system, the engagement of the public in the water management process, and the initiation and promotion of dialogue. Their system dynamics model consists of surface water and ground water budgets.

Researchers introduced a new approach called spatial system dynamics (SSD) to model feedback based dynamic processes in time and space. In this approach system dynamics and geographic information system (GIS) are coupled and it can be used to model a variety of physical and natural processes where the main interest is the space–time interaction. They applied the approach to flood management in the Red River basin in Manitoba, Canada.

Researchers applied a system dynamics modeling approach to improve the efficiency of water allocation incorporating a myriad of irrigation system constraints. They developed the Economical Reallocating Water Model (ERWM) based on an economic rationale, to analyze and determine the most economic water use under different water prices in the two hypothetical irrigation areas, which were constrained by environmental flow rules and pumping of ground water. ERWM consists of hydrological, agro-climatic and agro-economic components. The hydrological component represents the hydrological system in terms of surface water, ground water and hydrological constraints of pumping target and channel capacity; the agro-climatic component represents the cropping mix, crop area, evapotranspiration and rainfall; and the economic component that controls water price for each water source is used to link the hydrological and agro-climatic component. The model calculates the water used from each resource, final shortage, water cost and economic return to the region.

Researchers developed a complex system dynamic model (TianjinSD) for integrated and scientific management of the water resources of Tianjin, China. The model includes four major subsystems: population subsystem, economy subsystem, water pollution control subsystem and water resources subsystem. They verified the model using the data of 1996–2004 by historical examination of population, value added of industry, industrial water demand, value added of agriculture, and the total water demand in Tianjin. The model was applied for a 12 years period, taking 2004 as the base year and reasonable predictive results for policy-making on water resources allocation and management were achieved.

Researchers presented a dynamic simulation model of a water system in Yulin City and analyses the effectiveness of various supply and demand policies in meeting socio-economic and ecological requirements. Total water demand considered in the model includes production water demand, domestic water demand and ecological water demand; total water supply considered is a combination of surface water supply, groundwater supply, and other kinds of water supply (e.g. wastewater reuse and collection of rainwater). Implementation of water allocation is done by water pricing which are calculated combining resource price, engineering price and environment price. They examined three scenarios: businesses- usual, capacity acquisition and price-control and concluded that demand management represents the leverage point for achieving long-term sustainability outcomes.

Rainfall-Runoff Modeling

Rainfall-Runoff modeling is one of the most classical applications of hydrology. It has the purpose of simulating the peak river flow or the hydrograph induced by an observed or a hypothetical rainfall forcing. Rainfall-runoff models may include other input variables, like temperature, information on the catchment or others. We are studying rainfall-runoff models with the purpose of producing estimates of peak river flow, simulation of flood hydrographs or simulation of synthetic river flows in general, even for extended periods, for example for setting up water resources management strategies.

Rainfall-runoff models describe a portion of the water cycle and therefore the movement of a fluid - water - and therefore they are explicitly or implicitly based on the laws of physics, and in particular in the principles of conservation of mass, conservation of energy and conservation of momentum. Depending on their complexity, models can also simulate the dynamics of water quality, ecosystems, and other dynamical systems related to water, therefore embedding laws of chemistry, ecology, social sciences and so forth.

Models are built by constitutive equations, namely, mathematical formulations of the above laws, whose number depends on the number of variables to be simulated. The latter are the output variables, and the state variables, which one may need to, introduce to describe the state of the system. Constitutive equations may include parameters: they are numeric factors in the model equations that can assume different values therefore making the model flexible. In order to apply the model, parameters needs to be estimated (or calibrated, or optimized, and we say that the model is calibrated, parameterized, optimized). Parameters usually assume fixed value, but in some models they may depend on time, or the state of the system.

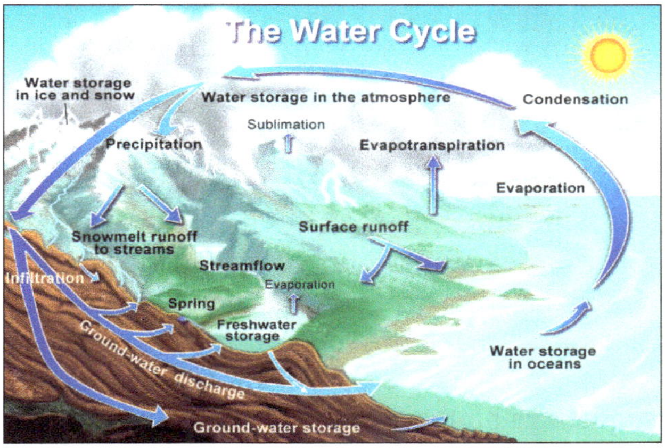

The hydrologic cycle.

Rainfall-runoff models can be classified within several different categories. They can distinguish between event-based and continuous-simulation models, black-box versus conceptual versus process based (or physically based) models, lumped versus distributed models, and several others. It is important to note that the above classifications are not rigid - sometimes a model cannot be unequivocally assigned to one category. We will treat rainfall-runoff models by taking into consideration models of increasing complexity.

The Rational Formula

It delivers an estimate of the peak river flow only, and that it is implicitly based on the principle of mass conservation. There is also the implicit use of the principle of conservation of energy in the estimation of the time of concentration. In fact, even if such time is estimated empirically, the mass transfer to the catchment outlet is actually governed by transformation and conservation of energy.

The Linear Reservoir

The linear reservoir is one of the most used rainfall-runoff models, together with the time area method. The linear reservoir model assimilates the catchment to a reservoir, for which the conservation of mass applies. The reservoir is fed by rainfall, and releases the river flow through a bottom discharge, for which a linear dependence applies between the river flow and the volume of water stored in the reservoir, while other losses - including evapotranspiration - are neglected. Therefore, the model is constituted by the following relationships:

$$\frac{dW(t)}{dt} = p(t) - q(t)$$

and

$$W(t) = kq(t)$$

where W(t) is the volume of water stored in the catchment at time t, p(t) is rainfall, q(t) is the river flow at time t and k is a constant parameter with the dimension of time (if the parameter was not constant the model would not be linear).

The second equation above assumes a linear relationship between discharge and storage into the catchment. Actually, the relationship between storage in a real tank and bottom discharge is an energy conservation equation that is not linear; in fact, as it is given by the well-known Torricelli's law. Therefore the linearity assumption is an approximation, which is equivalent to assuming that the superposition principle applies to runoff generation. Actually, such assumption does not hold in practice, as the catchment response induced by two subsequent rainfall events cannot be considered equivalent to the sum of the individual catchment responses to each single event. However, linearity is a convenient assumption to make the model simpler and analytical integration possible.

A nice feature of the linear reservoir is that the above equations can be integrated analytically, under simplifying assumption. In fact, by substituting the second equation into the first one gets:

$$k\frac{dq(t)}{dt} = p(t) - q(t)$$

Then, by multiplying both sides by $e^{t/k}$ and dividing by k one gets:

$$e^{t/k}\frac{dq(t)}{dt} + e^{t/k}\frac{q(t)}{k} - e^{t/k}\frac{p(t)}{k} = 0$$

which can be written as:

$$\frac{d}{dt}\left[e^{t/k}q(t)\right]=e^{t/k}\frac{p(t)}{k}$$

By integrating between 0 and t one obtains:

$$\int_0^t \frac{d}{d\tau}\left[e^{\tau/k}q(\tau)\right]d\tau=\int_0^t e^{\tau/k}\frac{p(\tau)}{k}d\tau$$

$$\left[e^{\tau/k}q(\tau)\right]_0^t=\int_0^t e^{\tau/k}\frac{p(\tau)}{k}d\tau$$

$$q(t)e^{t/k}-q(0)=\int_0^t e^{\tau/k}\frac{p(\tau)}{k}d\tau$$

and, by assuming q(0) = 0, one finally gets:

$$q(t)=\int_0^t \frac{1}{k}e^{\frac{-(t-\tau)}{k}}p(\tau)d\tau$$

which can be easily integrated numerically by using the Euler method.

If one assumes p(t) = constant, an explicit expression is readily obtained for the river flow:

$$q(t)=\frac{p}{k}\int_0^t e^{-\frac{t-\tau}{k}}d\tau=\frac{p}{k}e^{-\frac{t}{k}}\int_0^t e^{\frac{\tau}{k}}d\tau=\frac{p}{k}e^{-\frac{t}{k}}\left[ke^{\frac{\tau}{k}}\right]_0^t=\frac{p}{k}e^{-\frac{t}{k}}\left[ke^{\frac{t}{k}}-k\right]=p\left[1-e^{-\frac{t}{k}}\right]$$

The parameter k is usually calibrated by matching observed and simulated river flows. Its value significantly impacts the catchment response. A high k value implies a large storage into the catchment. Therefore, large values of k are appropriate for catchment with a significant storage capacity. Conversely, a low k value is appropriate for impervious basins. k is also related to the response time of the catchment. A low k implies a quick response, while slowly responding basins are characterised by a large k. In fact, k is related to the response time of the catchment.

The Non-Linear Reservoir

Several variants of the linear reservoir modeling scheme can be introduced, for instance by adopting a nonlinear relationship between discharge and storage. Moreover, an upper limit can be fixed for the storage in the catchment, and additional discharges can be introduced, which can be activated for different levels of storage. All of the above modifications make the model non-linear so that an analytical integration is generally not possible. One should also take into account that increasing the number of parameters implies a corresponding increase of estimation variance and therefore simulation uncertainty. Furthermore, it is often observed that introducing additional discharges or thresholds may induce discontinuities in the hydrograph shape.

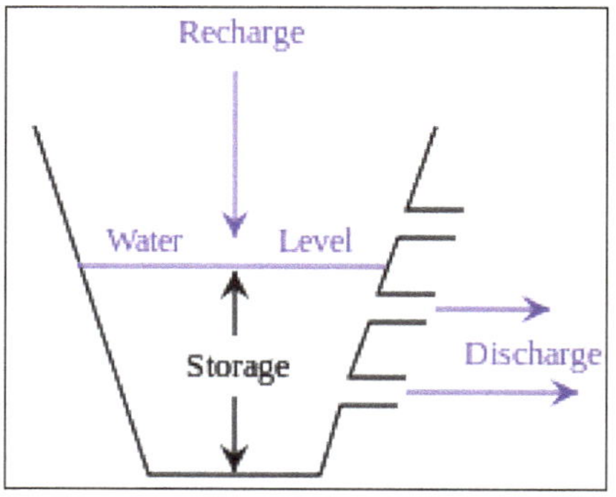

A non-linear reservoir.

The Hymod Model

The Hymod model is a flexible solution that is increasingly adopted for its capability of providing a good fit in several practical applications. It was originally proposed by Boyle. It is based on the assumption that each point location i in the basin is characterised by a local value of soil water storage C_i, which varies from 0 in the impervious areas up to a maximum value C_{max} in the most permeable location of the catchment. C_i is assumed to be randomly varying, so that for an assigned value C_* of soil water storage a probability distribution is introduced that gives the probability that a randomly selected location j is characterised by C_j less then, or equal to, C_*. Such probability may be interpreted as the fraction $F(C_*)$ of the catchment area where $C_j \le C_*$. The above probability distribution is written as,

$$F(C_*) = 1 - \left(1 - \frac{C_*}{C_{max}}\right)^{\beta_k}, \quad 0 \le C_* \le C_{max}.$$

Here, β_k is a parameter which quantifies the variability of the soil water storage over the catchment. One can easily verify by numerical simulation that $\beta_k=0$ implies that the soil water storage is constant over the basin and equal to C_{max}; $\beta_k=1$ implies that the soil water storage is linearly varying from 0 to C_{max}; $\beta_k \rightarrow \infty$ implies that the soil water storage is tending to the null value over the whole catchment, which is therefore impervious.

Let us assume that a storm event occurs over the basin and let us define with the symbol C(t) the time varying water depth stored in the unsaturated locations of the catchment. If we ignore any water losses, like evapotranspiration, C(t) is equal to the rainfall amount from the beginning of the event. If one assumes that the shape of the above probability distribution, now expressed in terms of C(t), is the one reported in figure, it can be easily proved that the water volume stored in the catchment at time t is given by,

$$W(t) = C(t) - \left[\int_0^{C(t)} F(\chi) d\chi\right].$$

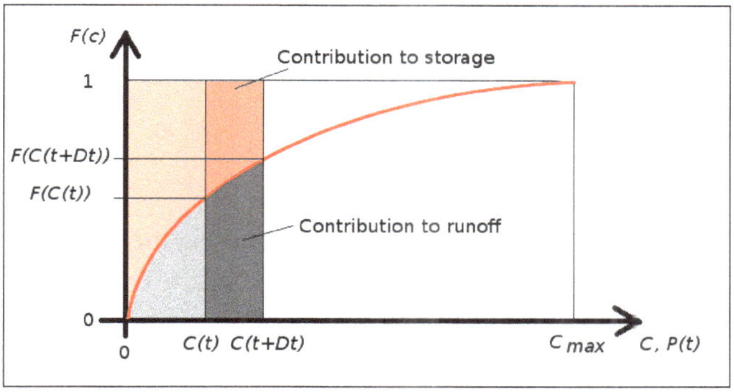

Distribution of soil water storage, surface runoff and stored volume in Hymod.

In fact, the integral at the right hand side of the above equation is the area below the red line in figure. Elementary increment of that area are given by the product of the rainfall at each time step by the fraction F(c(t)) of saturated area at the same time, namely, the product of rainfall by saturated area, which is indeed the surface runoff. Conversely, the area above the curve gives the global storage into the catchment W(t), which can be interpreted as a weighted average value of C(t). After saturation, the contribution of surface runoff is given by the product of the rainfall itself by 1, which is the fraction of saturated area, taking and keeping unit value when the catchment is saturated. After saturation, the storage in the catchment of course does not increase any more.

By computing the integral in the latter relationship one obtains:

$$\int_o^{C(t)} F(\chi)d\chi = \int_o^{C(t)} \left\{1-\left[1-\frac{\chi}{C_{max}}\right]^{\beta_k}\right\}d\chi,$$

$$\int_o^{C(t)} F(\chi)d\chi = \left[\chi+\frac{C_{max}}{(\beta_k+1)}\left[1-\frac{\chi}{C_{max}}\right]^{\beta_k+1}\right]_o^{C(t)},$$

$$\int_o^{C(t)} F(\chi)d\chi = C(t)+\frac{C_{max}}{(\beta_k+1)}\left[1-\frac{C(t)}{C_{max}}\right]^{\beta_k+1}-\frac{C_{max}}{(\beta_k+1)}.$$

Therefore the water volume stored in the catchment is equal to,

$$W(t)=\frac{C_{max}}{(\beta_k+1)}-\frac{C_{max}}{(\beta_k+1)}\left[1-\frac{C(t)}{C_{max}}\right]^{\beta_k+1}=\frac{C_{max}}{(\beta_k+1)}\left\{1-\left[1-\frac{C(t)}{C_{max}}\right]^{\beta_k+1}\right\}.$$

By inverting the above equation one gets,

$$C(t)=C_{max}\left\{1-\left[1-W(t)\frac{\beta_k+1}{C_{max}}\right]^{\frac{1}{\beta_k+1}}\right\}.$$

Note that one obtains an estimate of the upper value of the water volume that the catchment can store by imposing $C(t) = C_{max}$, therefore obtaining,

$$W_{max} = \frac{C_{max}}{(\beta_k + 1)}.$$

The above equations allow an easy application of the Hymod model through a numerical simulation, that is usually carried out by adopting a time step Dt that is equal to observational time step of rainfall and river flow. At a given time step t, one knows the value of C(t) which is equal to the cumulative rainfall depth from the beginning of the event at time t. Therefore, W(t) can be easily computed as well by using the above relationships. At the time t+1, C(t+1) = C(t)+P(t), where P is rainfall, under the condition that $C(t+1) = C_{max}$ if $C(t)+P(t)>C_{max}$. Therefore, one can compute a first contribution to surface runoff through the relationship,

$$ER_1(t) = \max\left(C(t)+P(t)-C_{max}, 0\right).$$

Finally, one can compute a second contribution to the surface runoff which is given by the water volume that cannot be absorbed by the catchment because part of the catchment area got saturated in the last time step. Such second contribution is given by,

$$ER_2(t) = \left(C(t+1)-C(t)\right)-\left(W(t+1)-W(t)\right).$$

By summing $ER_1 + ER_2$ one obtains the total contribution to the surface runoff during the time step from t to t+1.

At this stage, one may evaluate the water losses given by evapotranspiration at the current time step, and subtract them from W(t+1). Such water losses are computed within Hymod through the relationship,

$$E(t) = \left(1 - \frac{\frac{C_{max}}{\beta_k + 1} - W(t)}{\frac{C_{max}}{\beta_k + 1}}\right) E_P(t).$$

Here, $E_p(t)$ is the potential evapotranspiration at time t. Then, the water storage at time t+1 is given by,

$$W(t+1) = W(t) - E_p(t).$$

One should note that the evapotranspiration is subtracted from the stored water volume after ER_1 and ER_2 are computed.

The total contribution $ER(t) = ER_1(t) + ER_2(t)$ to the surface runoff is then divided into 2 components: aER(t) which represent the fast runoff and (1-a)ER(t) which is the slow runoff. aER(t) is propagated through a series of linear reservoirs with the same bottom discharge time constant k_q, while (1-a)ER(t) is instead propagated through a single linear reservoir with parameter k_s. The computation moves forward through the sequence of time steps. Hymod counts 5 parameters, namely: C_{max}, β, a, k_q and k_s. These parameters need to be calibrated by using observed data.

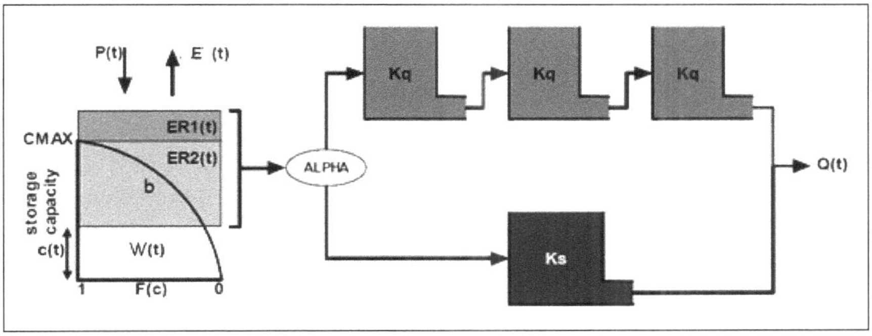

A schematic representation of the Hymod model.

References

- Water-resource-systems-modeling-its-role-in-Planning: eolss.net, Retrieved 16, February 2020
- A-Review-on-Water-Resources-Management-Modeling-272169299: researchgate.net, Retrieved 23, June 2020

3

Water Conservation and Water Integration System

Water conservation refers to the strategies and activities required to reduce unnecessary water usage to protect the hydrosphere. A few of the aspects related to water conservation are water harvesting, integrated water resources management and integrated urban water management. This chapter has been carefully written to provide an easy understanding of the varied facets of water conservation.

Water Conservation

Conservation is the philosophy that natural resources should be used cautiously so that they will remain available for future generations of people. In practice, conservation is the act of protecting, managing, and restoring shared earth resources such as soil, air, forests, minerals, petroleum, wildlife, and water—one of humans' most essential resources. Water conservation can be as simple as one person using water sparingly during a drought (prolonged period of dry weather), or as complex as a multi-national committee developing a long-term water distribution plan for an entire continent.

The word conservation means different things to different people, and a workable conservation plan for a particular region or resource usually involves a compromise between several interest groups. Consider, for example, a forest. To a logging company, forest conservation means developing a system of cutting and replanting healthy, fast-growing trees that ensures continuing profits. To a forest ecologist (a person who studies relationship between organisms and their environment) it means restoring a forest to a more natural state that supports a healthy community of plants and animals, along with protecting its most fragile areas and species. To a homeowner conservation means preserving the natural beauty of the forest and safeguarding property from forest fires. And to a preservationist it means letting nature manage the forest with little or no human intervention.

Conservation generally involves managing natural resources to serve people, whether by providing materials for essential needs (water, food, shelter, and energy) and consumer products (cars, clothing, computers, furniture), or simply by protecting wild areas where people can enjoy nature and outdoor recreation. While most scientists do not argue with using Earth's natural resources to meet human needs, most scientists also agree with the general idea that shared natural resources should be protected from overuse and pollution, and wisely managed using sound scientific information.

Need for Water Conservation

Water is Earth's most plentiful natural resource. It covers almost three-quarters of our planet's surface. Astronauts say Earth looks like a "blue marble" from space with its blue oceans and white swirls of water droplets (clouds) in the atmosphere (mass of air surrounding Earth). Water is also a renewable resource. Water moves endlessly within the hydrologic cycle, and is almost never destroyed in the process. When humans draw water from lakes, rivers, oceans, or groundwater reservoirs (aquifers), new water replaces it. Water is even self-cleansing. When it evaporates, it leaves pollution and salts behind and forms clouds that produce fresh rainwater. New water that flows into polluted lakes and rivers acts to dilute (lower the concentration or amount) and disperse (spread out) pollutants.

If water is so abundant and easily replenished, do people need to conserve it? Why then? First, fresh, liquid water that is suitable for human use only makes up a small percentage of Earth's total water supply. Almost all (97%) of Earth's water is undrinkable salt water that resides in the oceans. In addition to being too salty to drink, seawater corrodes (wears away) metals and gums up machinery, making it unsuitable for most other human uses. Much of the remaining 3% is frozen in glaciers (slow moving mass of ice) and ice in the North and South Poles. Second, fresh water is distributed unevenly on Earth's surface. Some regions have abundant freshwater resources and others are arid (dry) deserts where water is scarce. Third, though water is a renewable resource in a general sense, local and even regional water supplies can run dry from overuse. Finally, human activities that add chemical substances to surface water, groundwater and the oceans can pollute water to the extent that it is unsafe for human use and damages the larger ecosystem. For all these reasons, water conservation and management are extremely important, especially in places where a large human population depends on a limited water supply.

International Water Conservation

Although water shortages, floods, pollution, and water-related legal conflicts are relatively common in the United States, water conservation policies generally ensure that Americans can trust their water supply. People in other parts of the world are not so fortunate, particularly in the developing nations of Africa, South America, and Asia. In many regions arid climate, rapid population growth, poverty, and political instability are a recipe for water shortages and pollution. Two-thirds of the world's population lives on less than 13 gallons (49 liters) of water per day. (Remember that an average American uses about eight times that much water.) When political tension becomes war or an already dry climate gets drier, people who were surviving with limited freshwater are faced with famine (food shortages leading to starvation) and disease.

In recent decades conservation has become a critical issue for the international community. Organizations such as the United Nations Environment Program (UNEP), the International Union for the Conservation of Nature and Natural Resources (IUCN), and the World Wildlife Fund (WWF) are working to help individual countries plan for the maintenance and protection of their resources. Their strategy, called sustainable development, is based on a philosophy that is very similar to Pinchot's original conservation ideal. Earlier international programs viewed environmental protection and economic development as an "either-or" decision between preserving nature and human prosperity. Sustainable development schemes aim to address humans' most pressing social issues like poverty, famine, and disease by solving environmental problems such as water scarcity and pollution. New strategies for coping with environmental issues also involve providing economic incentives that encourage economically powerful nations and industries to act for the common good. The following methods would prove useful for it.

Protection of Water from Pollution

If the total fresh water available on the earth remains pollution free, it is sufficient to meet the drinking water needs of the existing population of the world, unfortunately a large portion of fresh water does not remain fit for use of the living world due to increasing economic activities, urbanization etc. Oceanic water in the form of ecological system of seas is an important environmental system, but during the last century pollution has spread in large proportions.

Surface water is mainly found in rivers and lakes and underground water is found under land at different depths, but which have become polluted. Large cities located on banks of rivers are directly disposing off different wastes without treatment in rivers. Similarly, tourism has spread pollution at warp speed on famous lakes and sea coasts. Man is greatly dependent on groundwater for his water related necessities, but some special industrial units have also polluted this amount of water stored in the security cover of the ground.

The entire available water store on the earth should be kept pollution-free because pressure of demand for water is increasing on a large part of the earth. Normally, it is presumed that sea water being saline, is not fit to be used by human beings, but indirectly it is useful for maintaining living organisms with whom man is related, e.g., fish provide nutrition to a large portion of the world.

Hence, it is necessary to have a cooperative policy at the international level for control of oily and radioactive pollution of the oceans. On 18th November, 2002, an oil tanker of Bahama met with an accident and drowned in the sea near the 'Coast of Death' sea area 233 km from north-west coast of Spain. 77,000 metric tons of oil poured out from this tanker and spread over the nearby seas and polluted the sea water. During the Gulf War of 2003 also, the water of Red Sea, Gulf Persian and Mediterranean Sea adjacent to the Arabian Sea became polluted.

The origin of water crisis initially started due to its pollution and qualitative deterioration, which became acute due to increasing demand for it. For preventing water pollution, important water sources should not be made places for disposal of wastes. Industrial units should dispose of water only after its treatment. Bathing and other such activities should be prohibited near drinking water sources.

Weeds produced in water should be controlled. Water should become re-usable after physical, chemical, mechanical and organic processing methods. Rivers and lakes should not be used as canals for transport of goods as is being done in Rhine River of Europe, Volga river of Russia, and the Ganges, Yamuna and Kaveri (Cauvery) rivers. Thus, the most important aspect of water conservation is control on water pollution.

Redistribution of Water

Water found on the surface of the earth is not equally distributed. Existing form of distribution also becomes a reason for the water crisis. In the African continent, though there is maximum hydroelectric production because of excess availability of water in Mediterranean regions, but the Sahara desert situated in the north of Africa and the greatest desert of the world, suffers from water crisis for the whole year.

Droughts faced by the Sahel region in the north of Sahara desert are world famous. Similarly, there is maximum rainfall of the world (1,187 cm. in Mawsynram) in north-east India, whereas there is only 50 cm. rainfall in the west. As a result of it, more than 60 per cent portion of water of Brahmaputra and its companion rivers in the north east flows to the saline seas uselessly, whereas rivers of western Rajasthan remain dry for most of the time in a year. Hence, by arranging supply of water from areas having lesser demand to the areas having greater demand, water crisis can be minimized. By construction of surface water reservoirs and storage of excess water in them, supply can be made to scarcity affected areas. This work can be accomplished by development of water reservoirs and canal network.

Excess rainfall water which flows away from rivers without being used, can be stored by construction of water reservoirs, from where it can be supplied for agriculture, industries, urban areas etc. Facilities of fisheries and transport also exist in stored water. These reservoirs are also constructed for protection from floods, and apart from flood protection, such water can also be used for different purposes. Redistribution of water is also possible through canal system. Canal system transfers water from excess rain water areas to scarce rain water areas and conserves water for different uses.

Rational Use of Groundwater

Groundwater meets 25 per cent of total supply of water in the world, remaining 75 per cent supply is met by surface water sources of rivers, lakes etc. Demand for groundwater goes on increasing in proportion to its available quantity due to which quantity of groundwater goes on decreasing. After exploitation of groundwater, its re-infiltration takes a very long time to complete.

Hence, groundwater exploitation should be only in proportion to its recharging capacity. The maximum groundwater is utilized for agriculture. Instead of growing crops requiring water according to agricultural climatic conditions, crops of commercial importance requiring more water are grown, due to which overexploitation of groundwater is done. By making only conscientious exploitation of groundwater, it can be proportionately conserved. For that, crops requiring lesser water should be given priority over crops requiring more water. Lesser water should be used by industries. For recharging groundwater, necessary drainage area should also be made available.

Population Control

Water crisis has assumed dangerous proportions due to fast increase in population and deterioration in quantitative and qualitative aspects of water resources. Demand for water is continuously increasing. Along with population increase, demand for fresh water has increased due to expansion of industries and increasing urbanization.

Demand for water increased 35 times from the year 1700 to the decade of 2000. Demand for water has increased more in developing countries. For the year 2001, demand for quantity of water in the whole world was estimated to be 435 cubic kms. 60 per cent of this is required in agriculture, 30 per cent in industries and 10 per cent for cooking, bathing and drinking. Hence, demand for water can also be controlled through population control and at the same time, qualitative deterioration of water can also be prevented by population. Population of the world would become around 800 crores by the year 2025. In this context, the United Nations Organization has warned that in case population control is not exercised, the whole world will have to face grave water crisis. United Nations Organization declared the year 2003 as the 'Fresh Water Year'.

Renovation of Traditional Water Sources

Traditional water storage places have been able to meet the demand of drinking water in many regions but they have been renovated from time to time. Water stored in traditional water sources has been used for both purposes, agriculture as well as for drinking. Important traditional water conservation methods for irrigation include Kuhul in hilly areas, Jing (Ladakh), Kool, Water Kundis called Khoop in Arunachal Pradesh, Zabo method of Nagaland, Aabi tanks of Haryana, Dong Pokhar of Assam, Bandhare of Maharashtra, Kere of Karnataka, Iree (Tanks) in Tamil Nadu, jackwell of Andaman

Nicobar and Nadi, Tanka, Kund, Khadeen, Kui, Bedi, Baori, Jhalra, Toba etc. of Rajasthan. Beginning of traditional water conservation methods was made in Java (Jordan) in the year 3,000 B.C. by construction of a vast water reservoir. A system of water conservation and drainage was found in Dhauliveera habitations, which was constructed during the Harappan civilization.

These traditional water conservation methods had been developed looking to the nature of rainfall in different regions, but the increasing population necessitated extension of agriculture, leading to deterioration of these traditional sources. Conservation of any traditional water source, instead of being limited to the water reservoir portion, extends to the whole drainage area system where rain water is stored after flowing.

Due to increasing area of agriculture, drainage area of water has been destroyed resulting in reduction of water flow therein. This has caused crisis for the existence of traditional water sources. Water crisis can be prevented by renovating traditional water conservation sources. Unless complete conservation of rain water is done in any region, the dream of water conservation cannot be fulfilled. Hence, supervision has to be done regarding renovation of dying traditional water conservation places. Now, the whole world has agreed with this fact that water available in its vast form also shall have to be conserved to keep it within our reach. From this point of view, ownership of traditional water sources should be at the village level as has also been agreed by the World Bank.

Use of Modern Irrigation Methods

At the world level, 69 per cent water is used for the agricultural sector. Necessary requirement of water in agriculture is met by surface water sources and groundwater. Surface water used for agriculture is obtained through canals and tanks and underground water from wells and tube wells. Unfortunately, much water is wasted through traditional methods of irrigation. Canal irrigation does not distribute water in a balanced way and it creates problems of water logging.

A large portion of water can be conserved by adoption of modern methods of irrigation. Irrigation consumes double the quantity of water in comparison to all other uses. Sprinkler and drip irrigation methods save 50 per cent water. In drop or drip irrigation method, pipes with holes are spread over the surface of land so that the crop directly receives water. There is no loss due to evaporation in this system and almost 95 per cent water is utilized. Thus, maximum water is utilized by this method. Improved modern irrigation methods are thus useful for conservation of water.

Increasing Forest Cover

According to hydrological movements, water is received through rainfall every year m different quantities on the surface of the earth. This water flows on the surface and reaches the seas. Some part of rainwater is stored in stable water reservoirs (lakes and tanks), whereas some quantity of water infiltrates into the land and takes the form of groundwater. Due to increasing deforestation during the last century, most of the rainwater flowed away to the saline seas without infiltrating into the ground. Water crisis also developed during the last decade in Cherapunji, which gets highest rainfall in the world, because forest cover has been destroyed there due to mining of limestone. As a result of it, rain water flows away very fast to the rivers.

The old tradition of tree plantation on the banks of rivers and tanks will have to be revived. Forest cover will have to be developed on uncultivable waste lands and hilly slopes on a large scale. Since trees bear drought conditions for a long duration as compared to crops, hence trees are helpful in reducing the demand for water along with recharging water sources.

Change in Crop Pattern

Excess water is not required if crops are grown according to agro-climatic conditions but in the present race of development, changing crop pattern with higher profits has replaced them. These commercial crops require more water than the traditional crops. In north-eastern part of Rajasthan, crops were not grown as per availability of water and intensive cultivation was adopted during the last three decades.

Groundwater was over exploited because of non-availability of surface water and plantation of commercial crops requiring more water. It created serious water crisis. Hence, keeping in mind the experience of Rajasthan, crop rotation should be adopted according to agro-climatic conditions. Agro forestry and horticulture should be given priority in areas having scarcity of water.

Flood Management

A large portion of fresh water in the world becomes devastating due to floods, out of a total land area of 32.8 crore hectares, 4 crore hectare land is flood affected, out of which 3.2 crore hectare land can be protected from floods. By construction of embankments and canals a large part of land can be conserved besides minimizing flood losses.

Intensive afforestation can also provide security from floods. If will be helpful in absorption of water in the soil. Drainage areas of Ganges, Yamuna, Mahanadi, Damodar, Kosi and other rivers have been taken up in flood management and security to some limit has been provided to the 1.44 crore hectare land.

Use of Geothermal Water

Water is also received from hot waterfalls regularly at many places on earth. Scarcity of water can be met to a certain extent by using such geothermal water.

Conserving Water in Industries

About 23 per cent of total fresh water available on the earth is used by industries the world over. Some special industries consume water to a large extent, whereas some industries pollute the major portion of water. Dyeing industry and leather industry are such type of industries which pollute the water. To manufacture one ton of steel, 300 tons of water is required. Consumption of water in industries is done in quantitative as well as qualitative forms. Proportionate consumption of water in industries in developed' countries is more (50%), out of which 75 per cent demand is met from surface water sources and 25 per cent from groundwater sources.

Water has to be both protected from pollution in industries, and re-used after processing. Recycling of water should be developed because normally industrial units dispose of water on the ground surface after using it only once. Because of not re-using it after processing, it also pollutes other water sources. Demand for water can also be reduced in industries by recycling used water.

Reuse of Urban Waste

Demand for water has increased in cities due to increasing urbanization. There is no provision for waste water treatment in many big towns and cities of various countries of the world. Instead of being reused, it pollutes other water sources.

Such condition is visible in cities of Delhi, Agra and Mathura on the banks of river Yamuna, whereas in many countries, urban water is used after treatment in nearby fields for growing vegetables and fruits. After use of water in urban areas, disposed waste water can be treated and conserved for use in agriculture in the peripheral areas of cities. Such policies should be incorporated while planning urban development.

Water Conservation by Municipal Bodies

Municipal bodies should manage both individual demand and supply of water as well as conserve water. Municipal laws should provide for collection of rain water from roof tops and implement it. Individual awareness is very important in water conservation. Every individual should develop the attitude of water conservation and prevent every drop of water from being wasted. As far as possible, one should conserve rain water along with maintaining quality of the water. Depth of traditional water sources should be maintained by controlling its cleanliness.

Since the maximum part of water is exploited in agriculture, hence water should be conserved in agriculture in different forms. Cultivating fields in off season helps in maintaining soil moisture. If cultivation is done up to 30 cm. depths, moisture can go up to 90 cm. depth, and capillary action prevents moisture coming out from the soil. Besides this, soil moisture can also be maintained by before-time sowing, proper use of fertilizers, pesticides and Weedicides. Green manure and crop rotation should also be adopted.

Environment balancing is the main basis of water conservation. Changes coming in the world environment result in shortage in the quantity of fresh water. Due to rise in temperature of the world, fresh water in the form of snow is melting and drifting towards saline oceans. This change is visible from Antarctica to Gangotri Glacier in the Himalayas.

Change in climate has also caused change in nature and quantity of rainfall, due to which existing unevenness in distribution of water has also increased. Thus, the mentality of World Environment Balancing has to be developed at the individual angle and conservation has to be worked on each stage. The objective behind the constitution of 'World Water Commission' at the Hague in 1999 was to consider questions related to water conservation and creating an environment so that strategy of water management may be worked out to search out solutions to the impending water crisis. Five groups have been constituted here.

First, for change in information system; second related to energy; third for considering issues related to organic technology; fourth to discuss institutional changes and the fifth panel is for consolidating serially future forecasts based on study of statistics. Apart from them, there is also a panel for traditional areas to consider water for food, water for human consumption and water for environment. One group also works on dams and national universality in international waters. The years 2025 and 2050 have been targeted by the commission for fulfilling its objective. The reason for such a long duration is that the results of solutions in this context are visible after a long time.

Water Harvesting

Water harvesting may be defined as "the process of concentrating precipitation through runoff and storage, for beneficial use". Water harvesting may occur naturally or by intervention. Besides being applied to agriculture, water harvesting may be developed to provide drinking water for humans and animals as well as for domestic and environmental purposes. Four main groups of water harvesting techniques can generally be distinguished: micro and macro-catchments, floodwater harvesting and storage reservoirs. Typical micro catchment techniques involve the delineation of natural depressions, the construction of contour and stone bunds, systems for inter-row water harvesting, terracing, construction of semicircular (half-moon) and triangular (V-shaped) bunds, eyebrow terraces, Vallerani-type micro-catchments, pits, muskets and negarim.

Rain Water Harvesting

Rainwater harvesting is a technology used for collecting and storing rainwater from rooftops, the land surface or rock catchments using simple techniques such as jars and pots as well as more complex techniques such as underground check dams. The techniques usually found in Asia and Africa arise from practices employed by ancient civilizations within these regions and still serve as a major source of drinking water supply in rural areas.

Benefits of Rainwater Harvesting

The collection of rainwater and harvesting it for everyday use has an ample number of exciting benefits. Let's take a look and discover some of these benefits.

Provides a Backup Source of Water

There are many uncertainties in water supply systems. Droughts may lead to a shortage of water. Destruction of water supply systems may occur at specific locations. Major repairs and maintenance in water companies happen sporadically. All these variables lead to the unavailability of water in your home.

So, installing a rainwater harvesting system is the best course of action to continue enjoying water supply in case any of the above variables occur. The collected rainwater can be used for many different purposes around the home. This includes laundry, washing dishes, and more. This would prevent groundwater depletion and augment groundwater table.

Ecological Benefits

The ecological benefits of rainwater harvesting are immense. An activity like flushing the toilet, for instance, attributes to about 25%-35% of home water usage. Add up the amount of water used to irrigate the lawn, do laundry, wash cars and the total amount of water used is simply astronomical. For businesses, more so hospitality establishments that accommodate a big number of guests in a single location, require a lot of water to satisfy customers. Rainwater harvesting will simply help to fill this gap.

The ecological benefit rainwater harvesting systems bring is that there will be a significant reduction in the amount of water used from the mains. This, ideally, means that water companies will not have to artificially pump in water from rivers and lakes. This will help minimize the possibility of rivers drying up. Another great ecological benefit of rainwater harvesting systems is that they help reduce the load placed upon drainage systems, minimizing the impacts of flooding by funneling the runoff water into large tanks for recycling.

Easy to Maintain

Rainwater harvesting systems are considerably easy to maintain since they are not utilized for drinking, cooking, or other sensitive uses. This means that they will not need the installation of purification systems, which are usually expensive. Rainwater harvesting systems are cost-effective, provide high-quality water, and reduce dependence on wells. The surplus rainwater can also be used to recharge groundwater aquifer through artificial recharge techniques.

Reduces Erosion and Flooding Around Buildings

Most buildings that utilize rainwater harvesting systems have a built-in catchment area on top of the roof, which is capable of collecting vast amounts of water in case of rainstorms. It helps to reduce soil erosion as by capturing rainwater, the flow of stormwater is reduced which can also prevent urban flooding. If the water were not collected efficiently, it could have resulted in serious soil erosion and flooding around the house.

Requires Low Upfront Capital Investment

Installation of a rainwater harvesting system is not that expensive, plus it's a one-time installation, and that's it. The only routine maintenance required is cleaning out the tank. You will probably hire an expert technician to install the system for you. The technician won't quote a high price, plus he will offer free advice on the best rainwater system to buy. He will also help out when buying the system to ensure you get the best quality rainwater harvesting system. When you sum up the overall cost of installing the system, you'll find that it's actually manageable.

Once you begin using rainwater you will also appreciate a reduced amount of water consumption in the home, thus all of your utility bills will greatly shrink in size. How incredible would it be to slash as much as 50 to 60% off of your water bill each month? No complaints about having additional money in the pocket each and every month.

Rainwater Harvesting Methods

It is simple to collect rainwater from building roofs and numerous other sources. As long as you are ready and equipped with a couple of different items you have all that it takes to begin rainwater harvesting and enjoying naturally delicious, clean, and useful water. Rainwater harvesting systems can be purchased complete from various home improvement stores. The cost of these systems will vary.

Rainwater Barrels

Barrel installation is the cheapest and easiest way to start harvesting rainwater in your home. A barrel is simply installed underneath the downspout of the guttering so that rain falling on the roof is funneled into the barrel. Usually, the container will consist of a spigot under it to enable you to draw water and irrigate your garden. Better still; a hose can be connected to the spigot or to a drip irrigation system.

Barrels are advantageous because they can effortlessly be sourced whether recycled or new. If you opt for the recycled barrel, make a point to know exactly what the barrel was used for in order to be sure that no chemical residues remain inside. If you reside in a mosquito-infested area, make sure to include a lid for your barrel to mitigate the possibility of mosquitoes breeding in your barrel.

When winter month knocks, it is advisable to disconnect your barrel to prevent freezing and cracking. The major disadvantage of rainwater barrels is that they have a limited capacity; meaning prolonged rains would result in flooding. The barrels are easy to use and readily available for purchase at many locations, all at reasonable prices.

Dry System

This is a well-known method of rainwater harvesting. It's basically an improvement to the rainwater barrel technique in terms of size. A larger water storage container is situated just a few steps from the property. This container provides a much larger storage capacity compared to the barrel. The gutter is then diverted to channel water into the tank. It's also easy to implement and relatively cheap. The name "dry" comes from the fact that the collection pipe dries up after any rainy day because it's meant to empty into the top of the tank. The dry system is ideal for areas with huge storms.

Wet System

The wet system is the exact opposite of the dry system. This is because the collection pipes are ever full of water since they are located underground. Here, numerous collection pipes are connected to multiple downspouts on the building and channeled to empty into the storage tank underground.

When there is no rain, the level stays constant, and the pipes are always full. Due to the ever availability of water inside the pipes, it is recommended that they be watertight to mitigate the possibility of leaking into the soil. The underground piping makes it relatively expensive to install compared to the dry and wet systems.

Green Roof

This method of rainwater harvesting literally cuts out the middleman. Instead of directing rainwater for storage in the tank and then channel it to the garden; you could install a green roof on your house to enable plants to use the water instantly. But you have to lay down a lining for your roof's protection and include a drainage system to take care of the runoffs.

Installing garden beds with plants that require low maintenance is the surefire way of maximizing your plot's productive space and leveraging the rainwater where it falls. Also, installing a green roof will add insulation to your house and cushion your roof from damage. If you prefer you can create your own barrels for rainwater harvesting. Just a few components like the Catchment area, conveyance system, first rain separator, filter unit, storage, delivery system, usage, recharge, and supplies are needed to create your own barrel.

You will need to follow the instructions included with your rainwater harvesting kit to set things up and begin the collection of the rainwater. For a DIY job, there are many resources available to you on the internet at no charge. You can definitely take advantage of the information whenever you need to and create your own rainwater harvesting system.

Integrated Water Resources Management

Integrated Water Resources Management (IWRM) is a process that promotes the coordinated development and management of water, land and related resources in order to maximize economic and social welfare in an equitable manner without compromising the sustainability of vital ecosystems.

IWRM is a cross-sectoral policy approach designed to replace the traditional, fragmented sectoral approach to water resources and management that has led to poor services and unsustainable resource use. Integrated Water Resources Management is based on the understanding that water resources are an integral component of the ecosystem, a natural resource, and a social and economic good.

The basis of IWRM is that the many different uses of finite water resources are interdependent. High irrigation demands and polluted drainage flows from agriculture, for example, mean less freshwater for drinking or industrial use; contaminated municipal and industrial wastewater pollutes rivers and threatens ecosystems; if water has to be left in a river to protect fisheries and ecosystems (environmental flows), less can be diverted to grow crops. There are plenty of other examples of the basic theme that unregulated use of scarce water resources is wasteful and inherently unsustainable.

IWRM is based on the three principles: social equity, economic efficiency and environmental sustainability. Considering these principles means answering the following questions:

- How will my decision/action affect access for other users to water or the benefits from its use?
- Will my decision/action result in the 'most efficient use of the available financial & water resources?
- How will my decision/action affect the functioning of natural systems?

Social equity means ensuring equal access for all users (particularly marginalised and poorer user groups) to an adequate quantity and quality of water necessary to sustain human well-being. The right of all users to the benefits gained from the use of water also needs to be considered when making water allocations. Benefits may include enjoyment of resources through recreational use or the financial benefits generated from the use of water for economic purposes.

Economic Efficiency means bringing the greatest benefit to the greatest number of users possible with the available financial and water resources. This requires that the most economically efficient option is selected. The economic value is not only about price – it should consider current and future social and environmental costs and benefits.

Ecological Sustainability requires that aquatic ecosystems are acknowledged as users and that adequate allocation is made to sustain their natural functioning. Achieving this criterion also requires that land uses and developments that negatively impact these systems are avoided or limited.

Operationally, IWRM approaches involve applying knowledge from various disciplines as well as the insights from diverse stakeholders to devise and implement efficient, equitable and sustainable solutions to water and development problems. As such, IWRM is a comprehensive, participatory planning and implementation tool for managing and developing water resources in a way that balances social and economic needs, and that ensures the protection of ecosystems for future generations. Water's many different uses—for agriculture, for healthy ecosystems, for people and livelihoods—demands coordinated action. An IWRM approach is an open, flexible process, bringing together decision-makers across the various sectors that impact water resources, and bringing all stakeholders to the table to set policy and make sound, balanced decisions in response to specific water challenges faced.

It has been agreed to consider water as a 'finite and economic commodity taking into account of affordability and equity criteria', in order to emphasize on its scarcity in the Dublin Statement:

- Fresh water is a finite and vulnerable resource, essential to sustain life, development and the environment.
- Water development and management should be based on a participatory approach, involving users, planners and policy makers at all levels.
- Women play a central part in the provision, management and safeguarding of water.
- Water has an economic value in all its competing uses and should be recognized as an economic good, taking into account of affordability and equity criteria.

Implementation of IWRM

IWRM aims to create sustainable water security within the present constraints and to improve the conditions in the catchment basin. Some important conditions for implementing IWRM are presented below:

- Political will and commitment: Political will at all levels can help unite all stakeholders and move the process forward. It is especially needed if the resulting plan or arrangement would create or require changes in legal and institutional structures, or if controversies and conflicts among stakeholders exist. Access to actors outside the water box is essential to move political will, gain sectoral support and ease public pressure for IWRM implementation.
- Basin management plan and clear vision: Water resources development coordinated among various sectors and users is facilitated by the preparation of a master plan that reflects the individual sector plans and offers the most effective and efficient utilization of the resource.
- Participation and coordination mechanisms, fostering information-sharing and exchange: The identification of key stakeholders can be facilitated through interviews and meetings. Stakeholder involvement can be defined appropriately for local conditions and improved gradually. Initial sharing of general basin-wide data and information, and further sharing of more specific information, will assist the self-sustaining system.
- Capacity development: Capacity development and training priorities should be expressed at all levels, including that of decentralized local government. Participants who may be adversely impacted and/or socially marginalized may be stimulated to participate within a consensus-building strategy.
- Well-defined flexible and enforceable legal frameworks and regulation: It is necessary to assemble and review the full range of existing laws and regulations

that apply to water-related activities and determine how existing legislation adapts or can be better adapted to accommodate sustainability and integration with regard to water resources management.

- Water allocation plans: As water is a shared resource, water rights should be flexible in terms of allocation in order to accommodate changes. Preparing a master plan that reflects individual sector plans facilitates the coordination among various sectors and advocates the most appropriate utilization of a basin's resource.
- Adequate investment, financial stability and sustainable cost recovery: Coordination for IWRM implementation needs financial sustainability – such as the promotion of cost recovery – and must consider long-term management. Various combinations and roles of international financing and donors such as government grants, public resources, user charges and taxes, donor funds, basin environmental trust funds can be considered as funding options.
- Good knowledge of the natural resources present in the basin: Adequate knowledge and information on the water resources inventory and human resources of the basin is desirable. Including scientists as water resource managers can help maintain and accrue sound knowledge of the natural resources.
- Comprehensive monitoring and evaluation: Monitoring and evaluation are essential for ensuring that the current management of water resources is properly implemented, and to identify the needs for adjusting management strategies. Upgrading new technologies is vital for effective performance both of local and central water management.

Integrated Urban Water Management

Integrated Urban Water Management (IUWM) calls for the alignment of urban development and basin management to achieve sustainable economic, social, and environmental goals. It brings together water supply, sanitation, storm and wastewater management and integrates these with land use planning and economic development.

An IUWM approach integrates planning for the water sector with other urban sectors, such as land, housing, energy, and transport to avoid fragmentation and duplication in policy- and decision-making. Cross-sector relationships are strengthened through a common working culture, collective goals and benefits are better articulated, and differences in power and resources can be negotiated. It includes the urban informal sector and marginalised communities.

IUWM principles are:

- Encompass alternative water sources.

- Match water quality with water use.
- Integrate water storage, distribution, treatment, recycling, and disposal.
- Protect, conserve and exploit water resources at their source.
- Account for non-urban users.
- Recognise and seek to align formal and informal institutions and practices.
- Recognise relationships among water, land use, and energy.
- Pursue efficiency, equity and sustainability.
- Encourage participation by all stakeholders.

The IUWM approach begins with clear national policies on integrated water resources management, backed by effective legislation to guide local authorities. A successful approach requires engaging local communities to solve the problems of water management. Collaborative approaches should involve all stakeholders in setting priorities, taking action, and assuming responsibility.

IUWM assesses both water quantity and quality, estimates future demand, anticipates the impacts of climate change, and recognises the importance of efficiency, without which water operations cannot be sustainable. It also recognises that different water sources can be used for different purposes – fresh water and desalinated water for domestic use; and treated wastewater for agriculture, industry, and the environment. With new desalination technologies, saltwater may become an accessible water source. IUWM requires the development of planning and management for all components of urban water services. Figure illustrates the coordinating structure that will ensure communication between departments, levels of government, local communities, and stakeholders.

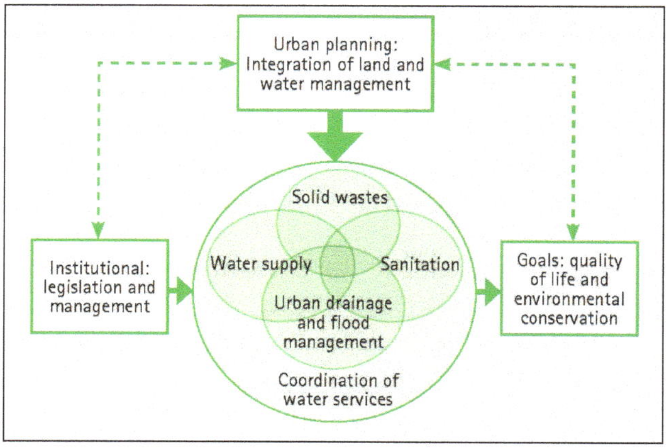

Urban planners can help governments overcome fragmented public policy and decision-making by linking planning with other policy sectors like infrastructure, and adopting collaborative approaches that involve all stakeholders in determining priorities, actions, and responsibilities. This may involve new ways of coordinating different agencies and controlling water use, such as new regulators to enforce standards and procedures. Integrated urban water policies based on participatory governance can secure sustainable development, but changes will be necessary to stimulate innovation, efficiency, and sustainability.

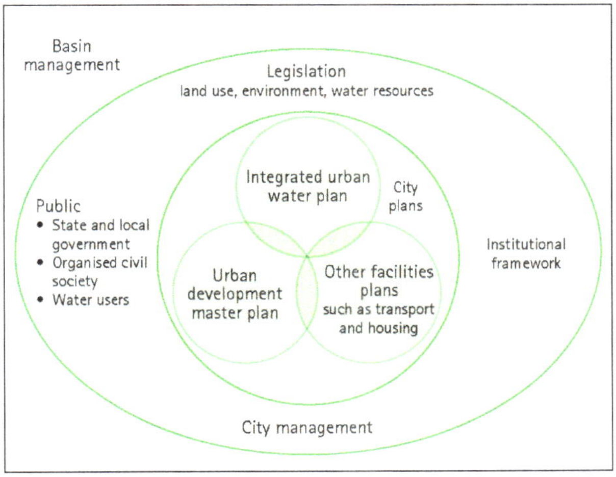

One constraint in adopting IUWM will be transforming entrenched institutional practice in large cities. Thus the greatest opportunities to achieve results may lie with small and medium-size cities of fewer than 500,000 inhabitants. Instilling a different approach to resource management in the governance of these cities is possible.

The Economics of IUWM

Under IUWM, water prices and allocations reflect the costs of developing and delivering water supplies and maintaining the system. Price signals the true value of water. Accurate pricing will encourage all users to manage water wisely, consistent with an integrated urban water management strategy. Differential tariffs that account for water quality can incentivize all users to reduce surface water or groundwater consumption in favour of reclaimed water for example. Tariffs, taxes, and subsidies can be used to distribute benefits fairly without diminishing the productivity of water resources. But if tariffs are set too low so they favour poor users and then cannot support effective operations and maintenance, the system may inadvertently contribute to greater inequality.

Pricing instruments can be designed so users pay more for higher levels of consumption or quality. Financial incentives like rebates, subsidised retrofits, water audits, and seasonal and zone pricing can also be used. Schemes under the 'polluter pays' principle, in which charges relate to the effluent that users generate, can improve the cost-effectiveness of treatment and reuse. They can even fund the construction of new infrastructure. But IUWM projects require significant levels of funding for both capital and operation and maintenance costs. For countries with limited ability to invest in water infrastructure, appropriate policies and well-functioning institutions are needed for fund-raising.

Policy Recommendations

Adopting IUWM and its iterative processes will help cities to significantly reduce the number of people without access to water and sanitation by providing water services of appropriate quantity and quality, and improving the health and productivity of urban residents. Governments should:

- Ensure their policies and strategies facilitate putting IUWM into practice at local and national levels, supported by financing strategies, technological developments, and tools for decision-making.
- Take on a more central role in cities and towns so as to lead development initiatives and ensure basic needs are met.
- Incorporate climate change predictions in planning urban water supply and sanitation and install and maintain, with the participation of a wide range of stakeholders, infrastructure and services that are 'climate-proof'.
- Pay special attention to supporting the informal urban sector, vital for a sustainable urban economy.
- Overcome governance fragmentation in public policy and decision-making by linking planning with the activities of other sectors.
- Build staff and institutional capacity to engage in IUWM to ensure they deliver at an optimal level.
- Engineer tariffs, taxes, and subsidies to transfer benefits to vulnerable groups, and ensure pricing policy reflects true costs.
- Consider employing the 'polluter pays' principle to improve the cost-effectiveness of treatment and reuse.

References

- 13-outstanding-water-conservation-methods-2435: biologydiscussion.com, Retrieved 05, May 2020
- Methods-of-rainwater-harvesting: conserve-energy-future.com, Retrieved 23, January 2020
- What-integrated-water-resources-management, integrated-water-Resources-Management-(I-WRM,the-sustainability-of-vital-ecosystems, where-we-work, disasters-conflicts: unenvironment.org, Retrieved 17, August 2020
- Integrated-water-resources-management-basic-concepts: iwapublishing.com, Retrieved 12, April 2020
- Water-conservation, scandinavian-literature-biographies, biographies, literature-and-arts: encyclopedia.com, Retrieved 28, July 2020

4

Water Pollution and its Types

The contamination of water bodies by harmful substances as a result of human activities is known as water pollution. Some of the topics studied under water pollution are groundwater pollution, flow dynamics and wastewater. The topics elaborated in this chapter will help in gaining a better perspective about water pollution.

Water Pollution

Raw sewage and industrial waste in the New River as it passes from Mexicali to Calexico, California.

Water pollution is the contamination of water bodies (e.g. lakes, rivers, oceans, aquifers and groundwater). This form of environmental degradation occurs when pollutants are directly or indirectly discharged into water bodies without adequate treatment to remove harmful compounds.

Water pollution affects the entire biosphere – plants and organisms living in these bodies of water. In almost all cases the effect is damaging not only to individual species and population, but also to the natural biological communities.

Water pollution is a major global problem which requires ongoing evaluation and revision of water resource policy at all levels (international down to individual aquifers and wells). It has been suggested that water pollution is the leading worldwide cause of deaths and diseases, and that it accounts for the deaths of more than 14,000 people daily. An estimated 580 people in India die of water pollution related illness every day. About 90 percent of the water in the cities of China is polluted. As of 2007, half a billion Chinese had no access to safe drinking water. In addition to the acute problems of water pollution in developing countries, developed countries also continue to struggle with pollution problems. For example, in the most recent national report on water quality in the United States, 44 percent of assessed stream miles, 64 percent of assessed lake acres, and 30 percent of assessed bays and estuarine square miles were classified as polluted. The head of China's national development agency said in 2007 that one quarter the length of China's seven main rivers were so poisoned the water harmed the skin.

Pollution in the Lachine Canal, Canada.

Water is typically referred to as polluted when it is impaired by anthropogenic contaminants and either does not support a human use, such as drinking water, or undergoes a marked shift in its ability to support its constituent biotic communities, such as fish. Natural phenomena such as volcanoes, algae blooms, storms, and earthquakes also cause major changes in water quality and the ecological status of water.

Categories

Although interrelated, surface water and groundwater have often been studied and managed as separate resources. Surface water seeps through the soil and becomes groundwater. Conversely, groundwater can also feed surface water sources. Sources of surface water pollution are generally grouped into two categories based on their origin.

Point Sources

Point source water pollution refers to contaminants that enter a waterway from a single, identifiable source, such as a pipe or ditch. Examples of sources in this category include discharges from a sewage treatment plant, a factory, or a city storm drain. The U.S. Clean Water Act (CWA) defines point source for regulatory enforcement purposes. The CWA definition of point source was amended in 1987 to include municipal storm sewer systems, as well as industrial storm water, such as from construction sites.

Point source pollution at a shipyard in Rio de Janeiro, Brazil.

Non-point Sources

Nonpoint source pollution refers to diffuse contamination that does not originate from a single discrete source. NPS pollution is often the cumulative effect of small amounts of contaminants gathered from a large area. A common example is the leaching out of nitrogen compounds from fertilized agricultural lands. Nutrient runoff in storm water from "sheet flow" over an agricultural field or a forest are also cited as examples of NPS pollution.

Blue drain and yellow fish symbol used by the UK Environment Agency to raise awareness of the ecological impacts of contaminating surface drainage.

Contaminated storm water washed off of parking lots, roads and highways, called urban runoff, is sometimes included under the category of NPS pollution. However, because this runoff is typically channeled into storm drain systems and discharged through pipes to local surface waters, it becomes a point source.

Groundwater Pollution

Interactions between groundwater and surface water are complex. Consequently, groundwater pollution, also referred to as groundwater contamination, is not as easily classified as surface water pollution. By its very nature, groundwater aquifers are susceptible to contamination from sources that may not directly affect surface water bodies, and the distinction of point vs. non-point source may be irrelevant. A spill or ongoing release of chemical or radionuclide contaminants into soil (located away from a surface water body) may not create point or non-point source pollution but can contaminate the aquifer below, creating a toxic plume. The movement of the plume, called a plume front, may be analyzed through a hydrological transport model or groundwater model. Analysis of groundwater contamination may focus on soil characteristics and site geology, hydrogeology, hydrology, and the nature of the contaminants.

Causes

The specific contaminants leading to pollution in water include a wide spectrum of chemicals, pathogens, and physical changes such as elevated temperature and discoloration. While many of the chemicals and substances that are regulated may be naturally occurring (calcium, sodium, iron, manganese, etc.) the concentration is often the key in determining what is a natural component of water and what is a contaminant. High concentrations of naturally occurring substances can have negative impacts on aquatic flora and fauna.

Oxygen-depleting substances may be natural materials such as plant matter (e.g. leaves and grass) as well as man-made chemicals. Other natural and anthropogenic substances may cause turbidity (cloudiness) which blocks light and disrupts plant growth, and clogs the gills of some fish species.

Many of the chemical substances are toxic. Pathogens can produce waterborne diseases in either human or animal hosts. Alteration of water's physical chemistry includes acidity (change in pH), electrical conductivity, temperature, and eutrophication. Eutrophication is an increase in the concentration of chemical nutrients in an ecosystem to an extent that increases the primary productivity of the ecosystem. Depending on the degree of eutrophication, subsequent negative environmental effects such as anoxia (oxygen depletion) and severe reductions in water quality may occur, affecting fish and other animal populations.

Pathogens

Disease-causing microorganisms are referred to as pathogens. Although the vast majority of bacteria are either harmless or beneficial, a few pathogenic bacteria can cause disease. Coliform bacteria, which are not an actual cause of disease, are commonly used as a bacterial indicator of water pollution. Other microorganisms sometimes found in surface waters that have caused human health problems include:

- *Burkholderia pseudomallei.*
- *Cryptosporidium parvum.*
- *Giardia lamblia.*
- *Salmonella.*
- *Norovirus* and other viruses.
- *Parasitic worms including the Schistosoma type.*

Poster to teach people in South Asia about human activities leading to the pollution of water sources.

A manhole cover unable to contain a sanitary sewer overflow.

Fecal sludge collected from pit latrines is dumped into a river at the Korogocho slum in Nairobi, Kenya.

High levels of pathogens may result from on-site sanitation systems (septic tanks, pit latrines) or inadequately treated sewage discharges. This can be caused by a sewage plant designed with less than secondary treatment (more typical in less-developed countries). In developed countries, older cities with aging infrastructure may have leaky sewage collection systems (pipes, pumps, valves), which can cause sanitary sewer overflows. Some cities also have combined sewers, which may discharge untreated sewage during rain storms.

Muddy river polluted by sediment.

Pathogen discharges may also be caused by poorly managed livestock operations.

Organic, Inorganic and Macroscopic Contaminants

Contaminants may include organic and inorganic substances.

A garbage collection boom in an urban-area stream in Auckland, New Zealand.

Organic water pollutants include:

- Detergents.
- Disinfection by-products found in chemically disinfected drinking water, such as chloroform.
- Food processing waste, which can include oxygen-demanding substances, fats and grease.

- Insecticides and herbicides, a huge range of organohalides and other chemical compounds.
- Petroleum hydrocarbons, including fuels (gasoline, diesel fuel, jet fuels, and fuel oil) and lubricants (motor oil), and fuel combustion byproducts, from storm water runoff.
- Volatile organic compounds, such as industrial solvents, from improper storage.
- Chlorinated solvents, which are dense non-aqueous phase liquids, may fall to the bottom of reservoirs, since they don't mix well with water and are denser.
    - Polychlorinated biphenyl (PCBs)
    - Trichloroethylene
- Perchlorate.
- Various chemical compounds found in personal hygiene and cosmetic products.
- Drug pollution involving pharmaceutical drugs and their metabolites.

Inorganic water pollutants include:

- Acidity caused by industrial discharges (especially sulfur dioxide from power plants).
- Ammonia from food processing waste.
- Chemical waste as industrial by-products.
- Fertilizers containing nutrients--nitrates and phosphates—which are found in storm water runoff from agriculture, as well as commercial and residential use.
- Heavy metals from motor vehicles (via urban storm water runoff) and acid mine drainage.
- Secretion of creosote preservative into the aquatic ecosystem.
- Silt (sediment) in runoff from construction sites, logging, slash and burn practices or land clearing sites.

Macroscopic pollution – large visible items polluting the water – may be termed "floatables" in an urban storm water context, or marine debris when found on the open seas, and can include such items as:

- Trash or garbage (e.g. paper, plastic, or food waste) discarded by people on the ground, along with accidental or intentional dumping of rubbish, that are washed by rainfall into storm drains and eventually discharged into surface waters.

- Nurdles, small ubiquitous waterborne plastic pellets.
- Shipwrecks, large derelict ships.

The Brayton Point Power Station in Massachusetts discharges heated water to Mount Hope Bay.

Thermal Pollution

Thermal pollution is the rise or fall in the temperature of a natural body of water caused by human influence. Thermal pollution, unlike chemical pollution, results in a change in the physical properties of water. A common cause of thermal pollution is the use of water as a coolant by power plants and industrial manufacturers. Elevated water temperatures decrease oxygen levels, which can kill fish and alter food chain composition, reduce species biodiversity, and foster invasion by new thermophilic species. Urban runoff may also elevate temperature in surface waters.

Thermal pollution can also be caused by the release of very cold water from the base of reservoirs into warmer rivers.

Transport and Chemical Reactions of Water Pollutants

Most water pollutants are eventually carried by rivers into the oceans. In some areas of the world the influence can be traced one hundred miles from the mouth by studies using hydrology transport models. Advanced computer models such as SWMM or the DSSAM Model have been used in many locations worldwide to examine the fate of pollutants in aquatic systems. Indicator filter-feeding species such as copepods have also been used to study pollutant fates in the New York Bight, for example. The highest toxin loads are not directly at the mouth of the Hudson River, but 100 km (62 mi) south, since several days are required for incorporation into planktonic tissue. The Hudson discharge flows south along the coast due to the coriolis force. Further south are areas of oxygen depletion caused by chemicals using up oxygen and by algae blooms, caused by excess nutrients from algal cell death and decomposition. Fish and shellfish kills have been reported, because toxins climb the food chain after small fish consume copepods, then large fish eat smaller fish, etc. Each successive step up the food chain causes a cumulative concentration of pollutants such as heavy metals (e.g. mercury) and persistent organic pollutants such as DDT. This is known as bio-magnification, which is occasionally used interchangeably with bio-accumulation.

A polluted river draining an abandoned copper mine on Anglesey.

Large gyres (vortexes) in the oceans trap floating plastic debris. The North Pacific Gyre, for example, has collected the so-called "Great Pacific Garbage Patch", which is now estimated to be one hundred times the size of Texas. Plastic debris can absorb toxic chemicals from ocean pollution, potentially poisoning any creature that eats it. Many of these long-lasting pieces wind up in the stomachs of marine birds and animals. This results in obstruction of digestive pathways, which leads to reduced appetite or even starvation.

Many chemicals undergo reactive decay or chemical change, especially over long periods of time in groundwater reservoirs. A noteworthy class of such chemicals is the chlorinated hydrocarbons such as trichloroethylene (used in industrial metal degreasing and electronics manufacturing) and tetrachloroethylene used in the dry cleaning industry. Both of these chemicals, which are carcinogens themselves, undergo partial decomposition reactions, leading to new hazardous chemicals (including dichloroethylene and vinyl chloride).

Groundwater pollution is much more difficult to abate than surface pollution because groundwater can move great distances through unseen aquifers. Non-porous aquifers such as clays partially purify water of bacteria by simple filtration (adsorption and absorption), dilution, and, in some cases, chemical reactions and biological activity; however, in some cases, the pollutants merely transform to soil contaminants. Groundwater that moves through open fractures and caverns is not filtered and can be transported as easily as surface water. In fact, this can be aggravated by the human tendency to use natural sinkholes as dumps in areas of karst topography.

There are a variety of secondary effects stemming not from the original pollutant, but a derivative condition. An example is silt-bearing surface runoff, which can inhibit the penetration of sunlight through the water column, hampering photosynthesis in aquatic plants.

Measurement

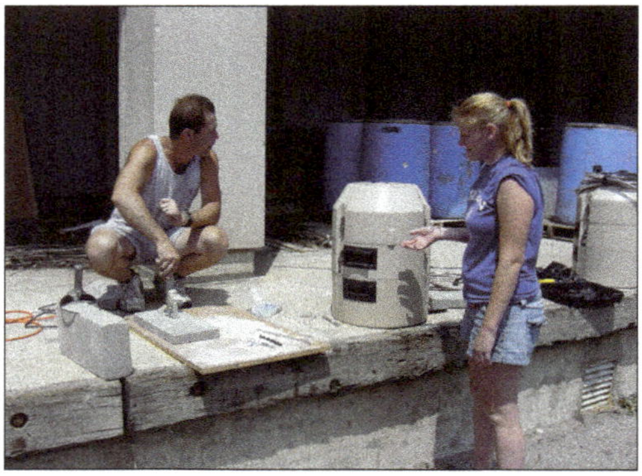

Environmental scientists preparing water autosamplers.

Water pollution may be analyzed through several broad categories of methods: physical, chemical and biological. Most involve collection of samples, followed by specialized analytical tests. Some methods may be conducted *in situ*, without sampling, such as temperature. Government agencies and research organizations have published standardized, validated analytical test methods to facilitate the comparability of results from disparate testing events.

Sampling

Sampling of water for physical or chemical testing can be done by several methods, depending on the accuracy needed and the characteristics of the contaminant. Many contamination events are sharply restricted in time, most commonly in association with rain events. For this reason "grab" samples are often inadequate for fully quantifying contaminant levels. Scientists gathering this type of data often employ auto-sampler devices that pump increments of water at either time or discharge intervals.

Sampling for biological testing involves collection of plants and/or animals from the surface water body. Depending on the type of assessment, the organisms may be identified for biosurveys (population counts) and returned to the water body, or they may be dissected for bioassays to determine toxicity.

Physical Testing

Common physical tests of water include temperature, solids concentrations (e.g., total suspended solids (TSS)) and turbidity.

Chemical Testing

Water samples may be examined using the principles of analytical chemistry. Many published test methods are available for both organic and inorganic compounds. Frequently used methods include pH, biochemical oxygen demand (BOD), chemical oxygen demand (COD), nutrients (nitrate and phosphorus compounds), metals (including copper, zinc, cadmium, lead and mercury), oil and grease, total petroleum hydrocarbons (TPH), and pesticides.

Biological Testing

Biological testing involves the use of plant, animal, and/or microbial indicators to monitor the health of an aquatic ecosystem. They are any biological species or group of species whose function, population, or status can reveal what degree of ecosystem or environmental integrity is present. One example of a group of bio-indicators are the copepods and other small water crustaceans that are present in many water bodies. Such organisms can be monitored for changes (biochemical, physiological, or behavioral) that may indicate a problem within their ecosystem.

Control of Pollution

Decisions on the type and degree of treatment and control of wastes, and the disposal and use of adequately treated wastewater, must be based on a consideration all the technical factors of each drainage basin, in order to prevent any further contamination or harm to the environment.

Sewage Treatment

Deer Island Wastewater Treatment Plant serving Boston, Massachusetts and vicinity.

In urban areas of developed countries, domestic sewage is typically treated by centralized sewage treatment plants. Well-designed and operated systems (i.e., secondary treatment or better) can remove 90 percent or more of the pollutant load in sewage. Some plants have additional systems to remove nutrients and pathogens.

Cities with sanitary sewer overflows or combined sewer overflows employ one or more engineering approaches to reduce discharges of untreated sewage, including:

- Utilizing a green infrastructure approach to improve storm water management capacity throughout the system, and reduce the hydraulic overloading of the treatment plant.

- Repair and replacement of leaking and malfunctioning equipment.
- Increasing overall hydraulic capacity of the sewage collection system (often a very expensive option).

A household or business not served by a municipal treatment plant may have an individual septic tank, which pre-treats the wastewater on site and infiltrates it into the soil.

Industrial Wastewater Treatment

Dissolved air flotation system for treating industrial wastewater.

Some industrial facilities generate ordinary domestic sewage that can be treated by municipal facilities. Industries that generate wastewater with high concentrations of conventional pollutants (e.g. oil and grease), toxic pollutants (e.g. heavy metals, volatile organic compounds) or other non-conventional pollutants such as ammonia, need specialized treatment systems. Some of these facilities can install a pre-treatment system to remove the toxic components, and then send the partially treated wastewater to the municipal system. Industries generating large volumes of wastewater typically operate their own complete on-site treatment systems. Some industries have been successful at redesigning their manufacturing processes to reduce or eliminate pollutants, through a process called pollution prevention.

Heated water generated by power plants or manufacturing plants may be controlled with:

- Cooling ponds, man-made bodies of water designed for cooling by evaporation, convection, and radiation.
- Cooling towers, which transfer waste heat to the atmosphere through evaporation and/or heat transfer.
- Cogeneration, a process where waste heat is recycled for domestic and/or industrial heating purposes.

Riparian buffer lining a creek in Iowa.

Agricultural Wastewater Treatment

Non Point Source Controls

Sediment (loose soil) washed off fields is the largest source of agricultural pollution in the United States. Farmers may utilize erosion controls to reduce runoff flows and retain soil on their fields. Common techniques include contour plowing, crop mulching, crop rotation, planting perennial crops and installing riparian buffers.

Nutrients (nitrogen and phosphorus) are typically applied to farmland as commercial fertilizer, animal manure, or spraying of municipal or industrial wastewater (effluent) or sludge. Nutrients may also enter runoff from crop residues, irrigation water, wildlife, and atmospheric deposition. Farmers can develop and implement nutrient management plans to reduce excess application of nutrients and reduce the potential for nutrient pollution.

To minimize pesticide impacts, farmers may use Integrated Pest Management (IPM) techniques (which can include biological pest control) to maintain control over pests, reduce reliance on chemical pesticides, and protect water quality.

Feedlot in the United States.

Point Source Wastewater Treatment

Farms with large livestock and poultry operations, such as factory farms, are called *concentrated animal feeding operations* or *feedlots* in the US and are being subject to increasing government regulation. Animal slurries are usually treated by containment in anaerobic lagoons before disposal by spray or trickle application to grassland. Constructed wetlands are sometimes used to facilitate treatment of animal wastes. Some animal slurries are treated by mixing with straw and composted at high temperature to produce a bacteriologically sterile and friable manure for soil improvement.

Erosion and Sediment Control from Construction Sites

Silt fence installed on a construction site.

Sediment from construction sites is managed by installation of:

- Erosion controls, such as mulching and hydroseeding, and
- Sediment controls, such as sediment basins and silt fences.

Discharge of toxic chemicals such as motor fuels and concrete washout is prevented by use of:

- Spill prevention and control plans, and
- Specially designed containers (e.g. for concrete washout) and structures such as overflow controls and diversion berms.

Control of Urban Runoff (Storm Water)

Effective control of urban runoff involves reducing the velocity and flow of storm water, as well as reducing pollutant discharges. Local governments use a variety of storm water management techniques to reduce the effects of urban runoff. These techniques, called best management practices (BMPs) in the U.S., may focus on water quantity control, while others focus on improving water quality, and some perform both functions.

Retention basin for controlling urban runoff.

Pollution prevention practices include low-impact development techniques, installation of green roofs and improved chemical handling (e.g. management of motor fuels & oil, fertilizers and pesticides). Runoff mitigation systems include infiltration basins, bioretention systems, constructed wetlands, retention basins and similar devices.

Thermal pollution from runoff can be controlled by storm water management facilities that absorb the runoff or direct it into groundwater, such as bioretention systems and infiltration basins. Retention basins tend to be less effective at reducing temperature, as the water may be heated by the sun before being discharged to a receiving stream.

United States Regulation of Point Source Water Pollution

Point source water pollution comes from discrete conveyances and alters the chemical, biological, and physical characteristics of water. In the United States, it is largely regulated by the Clean Water Act (CWA). Among other things, the Act requires dischargers to obtain a National Pollutant Discharge Elimination System (NPDES) permit to legally discharge pollutants into a water body. However, point source pollution remains an issue in some water bodies, due to some limitations of the Act. Consequently, other regulatory approaches have emerged, such as water quality trading and voluntary community-level efforts.

Definition

Water pollution is the contamination of natural water bodies by chemical, physical, radioactive or pathogenic microbial substances. Point sources of water pollution are described by the CWA as "any discernible, confined, and discrete conveyance from which pollutants are or may be discharged." These include pipes or man-made ditches from stationary locations such as sewage treatment plants, factories, industrial wastewater treatment facilities, septic systems, ships, and other sources that are clearly discharging pollutants into water sources.

Relevant Science

The input of pollutants into a water body may impact the water's ability to deliver ecological, recreational, educational, and economic services. While the impacts of water pollution vary considerably based on a variety of site-specific factors, they may be either direct or indirect. Pollutants that are directly toxic pose a threat to organisms that may come into contact with contaminated water. These include persistent organic pollutants used as pesticides and toxic byproducts of industrial activity (such as cyanide). Other pollutants may indirectly impact ecosystem services by causing a change in water conditions that allows for a harmful activity to take place. This includes sediment (loose soil) inputs that decrease the amount of light that can penetrate through the water, reducing plant growth and diminishing oxygen availability for other aquatic organisms.

There are a variety of water quality parameters that may be affected by point source water pollution. They include: dissolved oxygen and biochemical oxygen demand (BOD), temperature, pH, turbidity, phosphorus, nitrates, total suspended solids, conductivity, alkalinity, and fecal coliform. Given that much of the point source water pollution in the United States comes from municipal wastewater treatment plants, BOD is perhaps the most widely used metric to assess water quality.

Water quality is also closely linked with water quantity issues. Water shortages from natural and anthropogenic activity reduce the dilutive properties of water and may concentrate water pollution. Oppositely, during flooding events, water pollution may spread to previously uncontaminated waters through surface overflow or the failure of man-made barriers.

Nature of the Problem/Context

Cuyahoga River Fire

Before the CWA was enacted, companies indiscriminately discharged their effluents into water bodies. One such water body was the Cuyahoga River located in north-east Ohio. The river was thrust into the national limelight in 1969 when it caught fire, although the river had been plagued by fires since 1936. Pollution of the river had become prevalent in the early 1800s as contaminants from municipal and industrial discharges, bank erosion, commercial/residential development, atmospheric deposition, hazardous waste disposal sites, urban storm water runoff, combined sewer overflows (CSOs) and wastewater treatment plant bypasses were discharged into the river. *Time* magazine described the Cuyahoga as the river that "oozes rather than flows" and in which a person "does not drown but decays." The 1969 fire drew significant public attention to the state of the nation's waterways and is sometimes credited for the creation of the Federal Water Pollution Control Act (1972), commonly called the Clean Water Act, the Oil Pollution Act of 1990, and the establishment of federal and state agencies such as the Environmental Protection Agency (EPA).

Regulatory Framework

History of Regulation

Historically, regulation of point source water pollution in the United States included health- and use-based standards to protect environmental and economic interests. The Rivers and Harbors Act of 1899 contained provisions that made discharging refuse matter into navigable waters of the United States illegal without a permit issued by the U.S. Army Corps of Engineers.

In 1948 Congress passed the Federal Water Pollution Control Act (FWPCA). Although it was amended several times, the original FWPCA granted the Surgeon General of the Public Health Service the authority to develop programs to combat pollution that was harming surface and underground water sources. The FWPCA also authorized cooperation between federal and state agencies to construct waste treatment plants.

The FWPCA amendments of 1966, which came to be known as the *Clean Water Restoration Act* established a study to determine the effects of pollution on wildlife, recreation, and water supplies. The Act also set forth guidelines for abatement of water that may flow into international territory and prohibited the dumping of oil into navigable waters of the United States.

Further amendments to FWPCA in 1970 were dubbed the *Water Quality Improvement Act*. They required the development of certain water quality standards and expanded federal authority in upholding the standards. The most substantial amendments to the FWPCA occurred in 1972 and became known as the *Clean Water Act*.

Clean Water Act

Point source water pollution is largely regulated through the Clean Water Act, which gives the EPA the authority to set limits on the acceptable amount of pollutants that can be discharged into waters of the United States. The Act broadly defines a pollutant as any type of industrial, municipal, and agricultural waste discharged into water, such as: dredged soil, solid waste, incinerator residue, sewage, garbage, sewage sludge, munitions, chemical wastes, biological materials, radioactive materials, heat, wrecked or discarded equipment, rock, sand, cellar dirt and industrial, municipal, and agricultural waste. Point source water pollution is discharged into waters through both direct and indirect methods.

Indirect Point Source Water Pollution

An indirect discharger is one that sends its wastewater into a city sewer system, which carries it to the municipal sewage treatment plant or publicly owned treatment works (POTW). At the POTW, harmful pollutants in domestic sewage, called conventional pollutants, are removed from the sewage and then discharged into a surface water. POTWs are not designed to treat toxic or nonconventional pollutants in industrial wastewater, although they may incidentally remove some pollutants.

National Pretreatment Program

Indirect dischargers are covered by the National Pretreatment Program, which enforces three types of discharge standards:

- Prohibited discharge standards – protect against pass-through and interference.
- Categorical standards – national, uniform, technology-based standards that limit the discharge of pollutants.
- Local limits – address the specific needs of a POTW and its receiving waters.

The goal of the pretreatment program is to protect municipal wastewater treatment plants from damage that may occur when hazardous, toxic, or other wastes are discharged into a sewer system and to protect the quality of sludge generated by these plants. Discharges to a POTW are regulated primarily by the POTW itself, rather than the state/tribe or EPA.

Direct Point Source Water Pollution

Deer Island Waste Water Treatment Plant, serving the Boston, Massachusetts area.

Direct discharges are pollutants that are discharged directly into the water. To legally discharge pollutants directly into a waterbody, a National Pollution Discharge Elimination System (NPDES) permit must be obtained.

NPDES Permit Program

The NPDES permit program sets limits on the amount of pollutants that can be discharged into a waterbody. Technology based effluent limits establish a minimum level of pollution controls for all point source discharges. If technology based limits are not sufficient, water quality based effluent limits are developed.

Permits

Individual states are authorized by the EPA to issue permits when they have demonstrated that their program is at least as stringent as the EPA's program. States perform the day-to-day issuance of permits and oversight of the program while the EPA provides review and guidance to the states. All NPDES permits must contain a "specific, numeric, measurable set of limits on the amount of various pollutants that can appear in the wastewater discharged by the facility into the nation's waters" as well as guidelines on how often monitoring should be performed and what "sampling and analytic techniques should be used."

Types of Permits

- Individual – A unique permit is issued for each discharger.
- General – A single permit that covers a large number of similar dischargers in a specific geographic area. Examples include the EPA Vessels General Permit and industrial stormwater general permits.

Permitting Process

The authorized permit issuing body receives and reviews the permit application. The technology-based and water quality-based effluent limits are developed and then compared to determine which is the more stringent, which is then used as the effluent limit for the permit. Monitoring requirements, special conditions, and standard conditions for each pollutant are developed and the permit is then issued and its requirements are implemented.

- Technology-based requirements: A minimum level of treatment based on available treatment technologies is required for discharged pollutants, however, a discharger may use any available treatment technologies to meet the limits. The effluent limits are derived from different standards for different discharges:
    - Municipal discharges (POTW): National secondary treatment standards define limits of biological treatment standards based on biochemical oxygen demand (BOD), total suspended solids (TSS), and pH balance.
    - Non-municipal discharges: Limits for non-municipal (e.g. industrial) discharges are based on national discharge standards for industrial facilities within a certain category. These limits are achieved for using pollution control and prevention technologies designed for different types of dischargers. For existing dischargers, this level of treatment is equivalent to "Best Available Technology Economically Achievable" (BAT) and for new discharges, the treatment level is "New Source Performance Standards" (NSPS).
- Water quality-based requirements: Should technology-based standards not be stringent enough, Water Quality Based Effluent Limits (WQBEL) are developed to ensure that water quality standards are attained. WQBELs are based on ambient water quality standards.

Permit Components

All NPDES permits must contain the following five components:

- Cover page – indicates authorization for discharging and its locations.
- Effluent limits – limits used to control discharges through technology-based or water quality-based standards.
- Monitoring and reporting requirements – used to determine permit compliance.
- Special conditions – can be used to supplement effluent limits.
- Standard conditions – pre-established conditions that apply to all NPDES permits.

Permit Violation

A permitee can be in violation of their permit when they discharge pollutants at a level higher than what is specified on their permit or discharge without a permit. They can also be in violation if they fail to comply with the monitoring and enforcement portion of the permit.

Enforcement

Since the NPDES permit program is a self-monitoring system where permitees are required to carry out detailed monitoring requirements, the EPA promotes "compliance assistance" as an enforcement technique, which "helps permittees come into, and remain, in compliance with their permit, rather than going immediately to enforcement actions." The EPA and state NPDES agencies have can perform periodic inspections and the EPA gives individual states the authority to enforce NPDES permits although the EPA has the right to carry out enforcement should a state not do so. Enforcement actions for violations include: injunctions, fines for typical violations, imprisonment for criminal violations, or supplemental environmental projects (SEP). Citizens may also bring suits against violators but they must first provide the EPA and state NPDES permit agency with the opportunity to take action.

Stormwater Management Permits

To address the nationwide problem of stormwater pollution, Congress broadened the CWA definition of "point source" in 1987 to include industrial stormwater discharges and municipal separate storm sewer systems ("MS4"). These facilities were required to obtain NPDES permits. This 1987 expansion was promulgated in two phases. The Phase I regulation, promulgated in 1990, required that all municipalities of 100,000 persons or more, industrial dischargers, and construction sites of 5 acres (20,000 m^2) or more have NPDES permits for their stormwater discharges. Phase I permits were issued in much of the U.S. in 1991. The Phase II rule required that all municipalities, construction sites of 1 acre (4,000 m^2) or more, and other large property owners (such as school districts) have NPDES permits for their stormwater discharges. EPA published the Phase II regulation in 1999.

A silt fence, a type of sediment control, installed on a construction site.

Most construction sites are covered by general permits. Other industrial sites that only discharge stormwater are typically covered by general permits. Industrial stormwater dischargers that are otherwise required to have individual permits (due to their process wastewater and/or cooling water discharges), typically have the stormwater management requirements added to their individual permits.

In addition to implementing the NPDES requirements, many states and local governments have enacted their own stormwater management laws and ordinances, and some have published stormwater treatment design manuals. Some of these state and local requirements have expanded coverage beyond the federal requirements. For example, the State of Maryland requires erosion and sediment controls on construction sites of 5,000 sq ft (460 m^2) or more. It is not uncommon for state agencies to revise their requirements and impose them upon counties and cities; daily fines ranging as high as $25,000 can be imposed for failure to modify their local stormwater permitting for construction sites, for instance.

Others

EPA Water Quality Trading Policy

The Clean Water Act has made great strides in reducing point source water pollution, but this effect is overshadowed by the fact that nonpoint source pollution, which is not subject to regulation under the Act, has correspondingly increased. One of the solutions to address this imbalance is point/nonpoint source trading of pollutants. In January 2003, the EPA Water Quality Trading Policy was issued. At this time, many waters in the United States did not support their designated uses. Specifically, 40 percent of rivers, 45 percent of streams, and 50 percent of lakes that had been surveyed were unfit. Consequently, when The Water Quality Trading Policy was issued it acknowledged that "the progress made toward restoring and maintaining the chemical, physical, and biological integrity of the nation's waters under the 1972 Clean Water Act and its National Pollutant Discharge Elimination System (NPDES) permits has been incomplete."

The purpose of the policy is to "encourage voluntary trading programs that facilitate the implementation of TMDLs, reduce the costs of compliance with CWA regulations, establish incentives for voluntary reductions, and promote watershed-based initiatives (3)." The policy supports the trading of nutrients such as nitrogen and phosphorus and sediment load reductions, but in order for it to be extended to other contaminants, more scrutiny is required. All water quality trading programs are subject to the requirements of the Clean Water Act.

The Trading Policy outlines basic ground rules for trading by specifying viable pollutants, how to set baselines, and detailing the components of credible trading programs. It also stipulates that trades must occur within the same watershed. Water quality trading programs are subject to the stipulations of the Clean Water Act.

Other Laws that May Affect Some NPDES Permits

- Endangered Species Act: Federal agencies must consult with the U.S. Fish and Wildlife Service to ensure that discharges from a project (e.g., a new or expanded industrial facility) will not endanger a threatened species or their habitat.
- National Environmental Policy Act (NEPA): Only discharges that are subject to New Source Performance Standards (or new sources otherwise defined in the NPDES regulations) are subject to NEPA review prior to being issued a permit.
- National Historic Preservation Act: Prior to issuing a permit, EPA Regional Administrators must adopt measures that mitigate adverse effects on properties in the National Register of Historic Places.
- Coastal Zone Management Act: Permits will not be issued unless the permitees certifies that proposed activities, which would affect land or water use in coastal zones, comply with the Coastal Zone Management Act.
- Wild and Scenic Rivers Act: Prohibits issuance of permit for water resources projects that will have a direct, adverse effect on the values for which a national wild and scenic river was established.
- Fish and Wildlife Coordination Act: Jurisdiction over wildlife resources must be established prior to permit issuance so that resources can be conserved.
- *Magnuson-Stevens Act:* The "Essential Fish Habitat Provisions" in the law require EPA to consult with the National Marine Fisheries Service for any EPA-issued permits which may adversely affect essential fish habitat.

Problems/Issues/Concerns

Funding

Cost issues for monitoring

Monitoring of water bodies is the responsibility of authorized states, not the EPA. In 1997, EPA estimated that private and public point source control costs were $14 billion and $34 billion, respectively. The EPA has acknowledged that states have not adequately funded their monitoring programs, which has led to some uncertainty regarding the quality of most surface waters.

Enforcement

Self-monitoring and self-reporting

In many cases, the enforcement mechanisms of the Clean Water Act have created tension between regulators, regulated parties, and local citizens. Most NPDES permits require dischargers to monitor and report the contents of their discharges to the appropriate authorities. This requirement is potentially self-incriminating, forcing industries to provide information that may subject them to penalties and legal constraints. As a result, some dischargers go to great lengths to avoid penalties, including falsifying discharge monitoring reports and tampering with monitoring equipment. In *United States v. Hopkins*, the court ruled on a case where the vice president for manufacturing at Spirol International Corporation was charged with three criminal violations for falsifying water samples sent to state regulatory agencies. Spirol diluted his samples, which contained high levels of zinc, with tap water on numerous occasions and frequently ordered his subordinates to reduce the zinc concentration in the water by running it through a coffee filter.

Tensions between state and federal government

Like other environmental laws, the Clean Water Act delegates many enforcement responsibilities to state agencies. While the burden of enforcement may be transferred to the states, federal agencies reserve the right to approve or reject state plans for dealing with water pollution. This relationship reduces the regulatory burden on federal agencies, but can lead to confusion and tension between the two regulators.

Many of these tensions arise with regards to the commerce clause of the constitution. Until recently, the commerce clause has given the federal government considerable authority in regulating states' decisions about water use. In 2000, the United States Supreme Court ruled on *Solid Waste Agency of Northern Cook County v. US Army Corps of Engineers*. This ruling struck down the Corps' ability to prevent the construction of a disposal site for non-hazardous waste in Illinois based on power derived from the commerce clause. The Corps cited the Migratory Bird Rule when they initially denied the section 404 permit under the Clean Water Act. The migratory bird rule was meant to protect habitats used by migratory birds, which included the abandoned mining site that SWANCC had proposed to construct the waste disposal site. Chief Justice Rehnquist wrote: "Congress passes the CWA for the state purpose of 'restoring and maintaining the chemical, physical, and biological integrity of the Nation's waters.' In doing so,, Congress chose to recognize, preserve, and protect the primary responsibilities and rights of States to prevent, reduce, and eliminate pollution, to plan the development and use... of land and water resources...". In reversing the Corps' decision to issue a permit, the court reversed a trend and placed a check on federal power over state land use and water rights. Tensions between federal and state agencies concerning interstate commerce and point source water pollution continue, and are a reality of the Clean Water Act.

Ambiguity of the CWA

Coeur Alaska v. Southeast Alaska Conservation Council

In 2009, the Supreme Court ruled on *Coeur Alaska, Inc. v. Southeast Alaska Conservation Council.* The case concerned the re-opening of a gold mine outside Juneau, Alaska that had been out of operation since 1928. Coeur Alaska planned to utilize froth flotation in order to extract gold, creating 4.5 million tons of tailings over the course of its lifetime. The mining company opted to dispose of the tailings in nearby Lower Slate Lake, requiring a permit to comply with the Clean Water Act. The tailings would fundamentally change the physical and chemical characteristics of the lake, raising the lake bed by 50 feet and expanding the area from 23 to 60 acres. Coeur Alaska proposed to temporarily re-route nearby streams around Lower Slate Lake until they could purify the water and re-introduce the natural flow patterns.

Tailings from froth flotation contain high concentrations of heavy metals, including aluminum, which have toxic effects of aquatic organisms. As a result, the disposal of these tailings into Lower Slate Lake is eligible for a section 402 permit for discharge of a pollutant from the EPA (NPDES permit). The nature of the tailings also justifies their categorization as a fill material, or a "material [that] has the effect of... changing the bottom elevation" of a water body. Consequently, Coeur Alaska was also eligible for a "Dredge and Fill" permit from the Army Corps of Engineers under CWA section 404. The company applied for this latter permit and received authorization from the Corps to dump the tailings into Lower Slate Lake. The Southeast Alaska Conservation Council contended that disposal of the tailings is explicitly banned by section 306(e) of the Clean Water Act, and would therefore make Coeur Alaska ineligible for a NPDES permit.

The Court ruled in favor of Coeur Alaska, explaining that if the Army Corps of Engineers has authority to issue a permit under section 404, the EPA does not have authority to issue a section 402 permit. They asserted that the law is ambiguous as to whether section 306 applies to fill materials and found no erroneous or unreasonable behavior by the Corps. As a result, although the tailings would explicitly violate the Clean Water Act under section 402, the Corps may issue a dredge and fill permit.

This decision has not resonated well with environmental groups, who are worried that the decision may allow companies to discharge massive amounts of hazardous pollutants by avoiding the NPDES permitting procedure. Of particular concern is the mountaintop mining industry, which has the capacity to fundamentally alter aquatic ecosystems by filling in water bodies with sediment and mining debris. This tension between various sections within the Clean Water Act is sure to receive considerable attention in future years.

Other Emerging Regulatory Approaches

Water Quality Trading

Water quality trading (WQT) is a market-based approach, implemented on a watershed-scale, used to improve or maintain water quality. It involves the voluntary exchange of pollution reduction credits from sources with low costs of pollution control to those with high costs of pollution control. WQT programs are still subject to the requirements of the Clean Water Act, but they can be used to reduce the overall cost of compliance. Usually, permitted point sources of water pollution, such as wastewater treatment plants, have high discharge treatment costs, whereas nonpoint sources of water pollution, such as agriculture, have low costs of pollution reduction. Therefore, it is generally assumed that most trades would take place between point sources and nonpoint sources. However, point source-point source trades could also occur as well as pretreatment trades and intra-plant trades.

Most of the water quality trading markets currently in operation are focused on the trading of nutrients such as phosphorus and nitrogen. However, increasing interest has been shown in trading programs for sediment runoff, biological oxygen demand, and temperature. WQT programs can be used to preserve good water quality in unimpaired waters by counterbalancing new or increased pollutant discharges. In impaired waters, a WQT program can be used to improve water quality by reducing pollutant discharges in order to meet a specified water quality standard or total maximum daily load (TMDL).

TMDLs apply to both point sources and nonpoint sources and they represent the primary impetus for WQT programs. Point sources of pollutants that require NPDES permits often have strict discharge limits based on a TMDL. WQT can allow these sources to obtain lower costs of compliance, while still achieving the overall desired pollution reduction. Several factors influence whether or not a TMDL-based water quality trading program will be successful. First, the market must be appropriately structured within the regulatory framework of the Clean Water Act. Second, the pollutant must be well-suited for trading. Third, implementation of a WQT market requires public input and voluntary participation. Finally, there must be adequate differences in pollution control costs and available opportunities for reduction.

Credits and trade ratios

In a WQT market, a unit of pollutant reduction is called a credit. A point source can generate credits by reducing its discharge below its most stringent effluent limitation and a nonpoint source can generate credits "by installing best management practices (BMPs) beyond its baseline". Before being able to purchase credits, source must first meet its technology-based effluent limit (TBEL). The credits can then be used to meet water quality-based effluent limits (WQBEL).

In order to ensure that trades are effective and do not result in more pollution than would occur in their absence, trade ratios are used. Trade ratios can have several components including:

- Location: Source location relevant to the downstream area of concern can be an important factor.
- Delivery: The distance between sources can play a role in determining whether permit requirements are met at the outfall.
- Uncertainty: Nonpoint source reductions can be difficult to quantify.
- Equivalency: Sources may be discharging different forms of the same pollutant.
- Retirement: Credits may be retired to achieve further water quality improvement.

Permitted point sources can trade with other point sources or nonpoint sources. Trades can occur directly, or be brokered by third parties. However, when dealing with nonpoint source reductions, a level of uncertainty does exist. In order to address this, monitoring should be conducted. Modeling can also be used as a supplement to monitoring. Uncertainty can also be mitigated by field testing BMPs and using conservative assumptions for BMP efficacy.

Benefits

There are many economic, environmental, and social benefits that can be gained by establishing a WQT market within a watershed. Economically, since WQT is a market-based policy instrument, substantial savings can be generated while still achieving a mandated water quality goal. According to The National Cost to Implement Total Maximum Daily Loads (TMDLs) Draft report, flexible approaches to improving water quality, such as WQT markets, could save $900 million per year when compared with the least flexible approach (3). In 2008, WQT programs were worth $11 million, but have the potential for rapid growth. Other economic benefits of WQT include a reduction in the overall costs of compliance, the ability for dischargers to take advantages of economies of scale and differences in treatment efficiencies, and the ability to maintain growth without further harming the environment.

The environmental benefits of WQT programs are also numerous. First, habitats and ecosystems are protected and/or improved. Second, water quality objectives are able to be achieved in a timely manner. Third, there is incentive for innovation and creation of pollution prevention technologies. Finally, nonpoint sources are included in solving water quality problems. Social benefits include dialog among watershed stakeholders and incentives for all dischargers to reduce their pollution.

Factors influencing success

The success of a WQT market is determined by several factors including the pollutant of interest, physical characteristics of the affected watershed, control costs, trading mechanisms, and stakeholder participation and willingness. It is also important that the desired level of pollution reduction is not so great that all sources must reduce the maximum amount possible because this would eliminate surplus reductions to be used for credits.

Obstacles to implementation

The biggest obstacle to the widespread adoption of WQT markets is lack of supply and demand. In fact, studies of current water quality trading programs indicate that the typical problems associated with inhibition of water quality trading, such as high transaction costs, poor institutional infrastructure, and uncertain criteria, are being overcome. The main problem is that, under existing regulatory conditions, there are simply not enough willing buyers and sellers. Currently, most nonpoint sources of water pollution are unregulated or, assuming detection occurs, have relatively small consequences for violations. Consequently, nonpoint sources do not have incentive to participate in WQT. For WQT markets to be successful, greater demand is needed for pollution credits. For this to happen, water quality standards need to be clear and enforceable.

Contrast to emissions trading

Given the success of the sulfur dioxide emission trading market that was established to combat acid rain, at first glance it seems that this level of success should be easily extended to water quality trading. However, the reason this has not occurred yet comes down to a fundamental difference between water pollution and air pollution, and the process of establishing their respective markets. Establishing an emissions market, in principle, has three steps: (1) set a cap on emissions, (2) allocate portions of the cap to individual firms, and (3) allow each firm to meet its allowance through emission reduction or trade. The difference with water pollution, however, is that the problems that cause local water quality issues differ from those that create regional air pollution problems. Discharges into water are difficult to measure and effects are dependent on a variety of other factors and vary with weather and location.

Dealing with Transboundary Pollution: A Case Study of the U.S./ Mexico Border Region

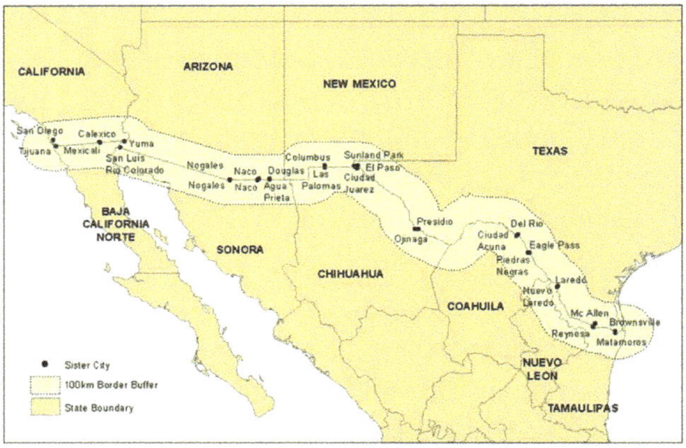

Map of the U.S. – Mexico border region.

Overview

The border region (approximately 2,000 miles (3,200 km) long and 62 miles (100 km) wide) is predominantly arid and contains seven watersheds including the Rio Grande which forms part of the border. The watersheds provide numerous benefits for the 14 cities and over nine million people in the region. However, increasing population, the arid climatic conditions of the region, the nature of economic activities along the rivers, increased trade, and uncontrolled emissions into them have placed tremendous pressure on water resources and threatened natural ecosystems. A large proportion of the population lacks access to clean drinking water and sanitation triggering public health concerns.

Policy Issues

Point source water pollution is a source of concern along the US-Mexico border as pollutants from both countries are entering shared waterways due to agricultural runoff, industrial discharge, and untreated sewage. Various policy issues arise in attempting to deal with this and include:

- Some pollution originates from areas beyond the border region as pollution is carried into the region by the waterways. This makes it difficult to regulate as discharges are difficult to apportion and control.
- Pollution is caused by and affects both countries therefore requiring a joint response.
- The socio-economic differences between the two countries have implications for policy implementation and enforcement.

- Various interests are represented with strong influence from environmental and social groups. Multiple levels of government agencies are also involved.

Policy Responses

There have been several attempts to address environmental concerns in the border region in the past by both governments. Significant intervention, however, resulted from the North American Free Trade Agreement (NAFTA) of 1994 between the U.S., Canada, and Mexico which renewed concerns over the environmental quality of the region due to increased trade in the region. The two governments therefore entered into the US-Mexico Border Environment Cooperation Agreement which created a number of institutions and programs. The Border Environmental Cooperation Commission(-BECC) and the North American Development Bank (NADB) were created to address border environmental-infrastructure issues and were effected in 2004. Distinct characteristics of these institutions and their approach are that they: are truly bi-national (have members from both countries) at all levels; emphasize a bottom-up approach with enhanced public participation; have a preference to assist disadvantaged communities; avoid regulatory or standard-driven approaches; emphasize economic and environmental sustainability.

The US-Mexico Border Program was also created by the agreement and placed under the management of the EPA (under regions 6 and 9) to correct the oversights of previous institutions and give guidance to cross-border environmental policy. The three institutions work together to identify, develop, finance and implement projects in the communities and certify them as "environmentally sustainable" subsequently funding them through a grant-making process. Communities, public, and private entities (sponsors) are invited to submit water and wastewater infrastructure projects. These projects are required to meet certain criteria to qualify for certification and funding. Among other requirements, they have to address an eligible environmental sector; must have a U.S.-side benefit; and have adequate planning, operations and maintenance, and pretreatment provisions. One specific provision touching on point-source water pollution states that "projects where the discharge is directly or indirectly to U.S. side waters, must target achievement of U.S. norms for ambient water quality in U.S. side waters, although infrastructure development may be phased over time. Any flow reductions that result from implementation of non-discharging alternatives must not threaten U.S. or shared ecosystems". Projects receive significant input from the communities living in the region in determining their sustainability. After certification, the project then receives funding from the NADB. So far, the BECC has certified 160 projects worth approximately three billion dollars since 2004.

The border program has also facilitated direct provision of infrastructure by the federal and state governments such as the construction of wastewater treatment plants, sewer lines, and raw water storage lagoons. One such example is the construction of the Matamoros lift station which is the first phase in eliminating raw sewage discharges into the Rio Grande. Further, the program emphasizes the provision of environmental education and information to communities living in the region. The EPA has established an informational website to provide news and information on the program.

Non-point Source Pollution

Muddy river.

Nonpoint source (NPS) pollution is a term used to describe pollution resulting from many diffuse sources, in direct contrast to point source pollution which results from a single source. Nonpoint source pollution generally results from land runoff, precipitation, atmospheric deposition, drainage, seepage, or hydrological modification (rainfall or snowmelt) where tracing the pollution back to a single source is difficult.

Non-point source water pollution affects a water body from sources such as polluted runoff from agricultural areas draining into a river, or wind-borne debris blowing out to sea. Non-point source air pollution affects air quality from sources such as smokestacks or car tailpipes. Although these pollutants have originated from a point source, the long-range transport ability and multiple sources of the pollutant make it a non-point source of pollution. Non-point source pollution can be contrasted with point source pollution, where discharges occur to a body of water or into the atmosphere at a single location.

NPS may derive from many different sources with no specific solution may change to rectify the problem, making it difficult to regulate. Non point source water pollution is difficult to control because it comes from the everyday activities of many different people, such as fertilizing a lawn, using a pesticide, or constructing a road or building.

It is the leading cause of water pollution in the United States today, with polluted runoff from agriculture the primary cause.

Other significant sources of runoff include hydrological and habitat modification, and silviculture (forestry).

Contaminated stormwater washed off parking lots, roads and highways, and lawns (often containing fertilizers and pesticides) is called urban runoff. This runoff is often classified as a type of NPS pollution. Some people may also consider it a point source because many times it is channeled into municipal storm drain systems and discharged through pipes to nearby surface waters. However, not all urban runoff flows through storm drain systems before entering water bodies. Some may flow directly into water bodies, especially in developing and suburban areas. Also, unlike other types of point sources, such as industrial discharges, sewage treatment plants and other operations, pollution in urban runoff cannot be attributed to one activity or even group of activities. Therefore, because it is not caused by an easily identified and regulated activity, urban runoff pollution sources are also often treated as true non-point sources as municipalities work to abate them.

Principal Types

Runoff of soil and fertilizer during a rain storm.

Sediment

Sediment (loose soil) includes silt (fine particles) and suspended solids (larger particles). Sediment may enter surface waters from eroding stream banks, and from surface runoff due to improper plant cover on urban and rural land. Sediment creates turbidity (cloudiness) in water bodies, reducing the amount of light reaching lower depths, which can inhibit growth of submerged aquatic plants and consequently affect species which are dependent on them, such as fish and shellfish. High turbidity levels also inhibit drinking water purification systems.

Sediment can also be discharged from multiple different sources. Sources include construction sites (although these are point sources, which can be managed with erosion controls and sediment controls), agricultural fields, stream banks, and highly disturbed areas.

Nutrients

Nutrients mainly refers to inorganic matter from runoff, landfills, livestock operations and crop lands. The two primary nutrients of concern are phosphorus and nitrogen.

Phosphorus is a nutrient that occurs in many forms that are bioavailable. It is notoriously over-abundant in human sewage sludge. It is a main ingredient in many fertilizers used for agriculture as well as on residential and commercial properties, and may become a limiting nutrient in freshwater systems and some estuaries. Phosphorus is most often transported to water bodies via soil erosion because many forms of phosphorus tend to be adsorbed on to soil particles. Excess amounts of phosphorus in aquatic systems (particularly freshwater lakes, reservoirs, and ponds) leads to proliferation of microscopic algae called phytoplankton. The increase of organic matter supply due to the excessive growth of the phytoplankton is called eutrophication. A common symptom of eutrophication is algae blooms that can produce unsightly surface scums, shade out beneficial types of plants, and poison the water due to toxins produced by the algae. These toxins are a particular problem in systems used for drinking water because some toxins can cause human illness and removal of the toxins is difficult and expensive. Bacterial decomposition of algal blooms consumes dissolved oxygen in the water, generating hypoxia with detrimental consequences for fish and aquatic invertebrates.

Nitrogen is the other key ingredient in fertilizers, and it generally becomes a pollutant in saltwater or brackish estuarine systems where nitrogen is a limiting nutrient. Similar to phosphorus in fresh-waters, excess amounts of bioavailable nitrogen in marine systems lead to eutrophication and algae blooms. Hypoxia is an increasingly common result of eutrophication in marine systems and can impact large areas of estuaries, bays, and near shore coastal waters. Each summer, hypoxic conditions form in bottom waters where the Mississippi River enters the Gulf of Mexico. During recent summers, the aereal extent of this "dead zone" is comparable to the area of New Jersey and has major detrimental consequences for fisheries in the region.

Nitrogen is most often transported by water as nitrate (NO_3). The nitrogen is usually added to a watershed as organic-N or ammonia (NH_3), so nitrogen stays attached to the soil until oxidation converts it into nitrate. Since the nitrate is generally already incorporated into the soil, the water traveling through the soil (i.e., interflow and tile drainage) is the most likely to transport it, rather than surface runoff.

Toxic Contaminants and Chemicals

Compounds including heavy metals like lead, mercury, zinc, and cadmium, organics like polychlorinated biphenyls (PCBs) and polycyclic aromatic hydrocarbons (PAHs), fire retardants, and other substances are resistant to breakdown. These contaminants can come from a variety of sources including human sewage sludge, mining operations, vehicle emissions, fossil fuel combustion, urban runoff, industrial operations and landfills.

Toxic chemicals mainly include organic compounds and inorganic compounds. These compounds include pesticides like DDT, acids, and salts that have severe effects to the ecosystem and water-bodies. These compounds can threaten the health of both humans and aquatic species while being resistant to environmental breakdown, thus allowing them to persist in the environment. These toxic chemicals could come from croplands, nurseries, orchards, building sites, gardens, lawns and landfills.

Acids and salts mainly are inorganic pollutants from irrigated lands, mining operations, urban runoff, industrial sites and landfills.

Pathogens

Pathogens are bacteria and viruses that can be found in water and cause diseases in humans. Typically, pathogens cause disease when they are present in public drinking water supplies. Pathogens found in contaminated runoff may include:

- *Cryptosporidium parvum*
- *Giardia lamblia*
- *Salmonella*
- *Novovirus* and other viruses
- Parasitic worms (helminths)

Coliform bacteria and fecal matter may also be detected in runoff. These bacteria are a commonly used indicator of water pollution, but not an actual cause of disease.

Pathogens may contaminate runoff due to poorly managed livestock operations, faulty septic systems, improper handling of pet waste, the over application of human sewage sludge, contaminated storm sewers, and sanitary sewer overflows.

Principal Sources

Urban and Suburban Areas

Urban and suburban areas are a main sources of nonpoint source pollution due to the amount of runoff that is produced due to the large amount of paved surfaces. Paved surfaces, such as asphalt and concrete are impervious to water penetrating them. Any water that is on contact with these surfaces will run off and be absorbed by the surrounding environment. These surfaces make it easier for stormwater to carry pollutants into the surrounding soil.

Construction sites tend to have disturbed soil that is easily eroded by precipitation like rain, snow, and hail. Additionally, discarded debris on the site can be carried away by runoff waters and enter the aquatic environment.

Typically, in suburban areas, chemicals are used for lawn care. These chemicals can end up in runoff and enter the surrounding environment via storm drains in the city.

Since the water in storm drains is not treated before flowing into surrounding water bodies, the chemicals enter the water directly.

Agricultural Operations

Agricultural operations account for a large percentage of all nonpoint source pollution in the United States. When large tracts of land are plowed to grow crops, it exposes and loosens soil that was once buried. This makes the exposed soil more vulnerable to erosion during rainstorms. It also can increase the amount of fertilizer and pesticides carried into nearby bodies of water.

Atmospheric Inputs

Atmospheric inputs of pollution into the air can come from multiple sources. Typically, industrial facilities, like factories, emit air pollution via a smokestack. Although this is a point source, due to the distributional nature, long-range transport, and multiple sources of the pollution, it is considered a nonpoint source. Additionally, atmospheric pollution can become water pollution, by being washed out of the atmosphere in the form of rain or snow.

Highway Runoff

Highway runoff accounts for a small but widespread percentage of all nonpoint source pollution. Harned (1988) estimated that runoff loads were composed of atmospheric fallout (9%), vehicle deposition (25%) and highway maintenance materials (67%) he also estimated that about 9 percent of these loads were reentrained in the atmosphere.

Forestry and Mining Operations

Forestry and mining operations can have significant inputs to non-point source pollution.

Forestry

Forestry operations reduce the number of trees in a given area, thus reducing the oxygen levels in that area as well. This action, coupled with the heavy machinery rolling over the soil increases the risk of erosion.

Mining

Active mining operations are considered point sources, however runoff from abandoned mining operations contribute to nonpoint source pollution. In strip mining operations, the top of the mountain is removed to expose the desired ore. If this area is not properly reclaimed once the mining has finished, soil erosion can occur. Additionally, there can be chemical reactions with the air and newly exposed rock to create acidic runoff.

Water that seeps out of abandoned subsurface mines can also be highly acidic. This can seep into the nearest body of water and change the pH in the aquatic environment.

Marinas and Boating Activities

Chemicals used for boat maintenance, like paint, solvents, and oils find their way into water through runoff. Additionally, spilling fuels or leaking fuels directly into the water from boats contribute to nonpoint source pollution. Nutrient and bacteria levels are increased by poorly maintained sanitary waste receptacles on the boat and pump-out stations.

Control

Contour buffer strips used to retain soil and reduce erosion.

Urban and Suburban Areas

To control non-point source pollution, many different approaches can be undertaken in both urban and suburban areas. Buffer strips provide a barrier of grass in between impervious paving material like parking lots and roads, and the closest body of water. This allows the soil to absorb any pollution before it enters the local aquatic system. Retention ponds can be built in drainage areas to create an aquatic buffer between runoff pollution and the aquatic environment. Runoff and storm water drain into the retention pond allowing for the contaminants to settle out and become trapped in the pond. The use of porous pavement allows for rain and storm water to drain into the ground beneath the pavement, reducing the amount of runoff that drains directly into the water body. Restoration methods such as constructing wetlands are also used to slow runoff as well as absorb contamination.

Construction sites typically implement simple measures to reduce pollution and runoff. Firstly, sediment or silt fences are erected around construction sites to reduce the amount of sediment and large material draining into the nearby water body. Secondly, laying grass or straw along the border of construction sites also work to reduce non-point source pollution.

In areas served by single-home septic systems, local government regulations can force septic system maintenance to ensure compliance with water quality standards. In

Washington (state), a novel approach was developed through a creation of a "shellfish protection district" when either a commercial or recreational shellfish bed is downgraded because of ongoing nonpoint source pollution. The shellfish protection district is a geographic area designated by a county to protect water quality and tideland resources, and provides a mechanism to generate local funds for water quality services to control nonpoint sources of pollution. At least two shellfish protection districts in south Puget Sound have instituted septic system operation and maintenance requirements with program fees tied directly to property taxes.

Agricultural operations

To control sediment and runoff, farmers may utilize erosion controls to reduce runoff flows and retain soil on their fields. Common techniques include contour plowing, crop mulching, crop rotation, planting perennial crops and installing riparian buffers. *Conservation tillage* is a concept used to reduce runoff while planting a new crop. The farmer leaves some crop reside from the previous planting in the ground to help prevent runoff during the planting process.

Nutrients are typically applied to farmland as commercial fertilizer; animal manure; or spraying of municipal or industrial wastewater (effluent) or sludge. Nutrients may also enter runoff from crop residues, irrigation water, wildlife, and atmospheric deposition. Farmers can develop and implement nutrient management plans to reduce excess application of nutrients.

To minimize pesticide impacts, farmers may use Integrated Pest Management (IPM) techniques (which can include biological pest control) to maintain control over pests, reduce reliance on chemical pesticides, and protect water quality.

Forestry Operations

With a well-planned placement of both logging trails, also called skid trails, can reduce the amount of sediment generated. By planning the trails location as far away from the logging activity as possible as well as contouring the trails with the land, it can reduce the amount of loose sediment in the runoff. Additionally, by replanting trees on the land after logging, it provides a structure for the soil to regain stability as well as replaces the logged environment.

Marinas

Installing shut off valves on fuel pumps at a marina dock can help reduce the amount of spillover into the water. Additionally, pump-out stations that are easily accessible to boaters in a marina can provide a clean place in which to dispose of sanitary waste without dumping it directly into the water. Finally, something as simple as having trash containers around a marina can prevent larger objects entering the water.

Groundwater Pollution

Groundwater pollution example in Lusaka, Zambia where the pit latrine in the background is polluting the shallow well in the foreground with pathogens and nitrate.

Groundwater pollution (also called groundwater contamination) occurs when pollutants are released to the ground and make their way down into groundwater. It can also occur naturally due to the presence of a minor and unwanted constituent, contaminant or impurity in the groundwater, in which case it is more likely referred to as contamination rather than pollution.

The pollutant creates a contaminant plume within an aquifer. Movement of water and dispersion within the aquifer spreads the pollutant over a wider area. Its advancing boundary, often called a plume edge, can intersect with groundwater wells or daylight into surface water such as seeps and spring, making the water supplies unsafe for humans and wildlife. The movement of the plume, called a plume front, may be analyzed through a hydrological transport model or groundwater model. Analysis of groundwater pollution may focus on soil characteristics and site geology, hydrogeology, hydrology, and the nature of the contaminants.

Pollution can occur from on-site sanitation systems, landfills, effluent from wastewater treatment plants, leaking sewers, petrol filling stations or from over application of fertilizers in agriculture. Pollution (or contamination) can also occur from naturally occurring contaminants, such as arsenic or fluoride. Using polluted groundwater causes hazards to public health through poisoning or the spread of disease.

Different mechanisms have influence on the transport of pollutants, e.g. diffusion, adsorption, precipitation, decay, in the groundwater. The interaction of groundwater contamination with surface waters is analyzed by use of hydrology transport models.

Pollutant Types

Contaminants found in groundwater cover a broad range of physical, inorganic chemical, organic chemical, bacteriological, and radioactive parameters. Principally, many of the same pollutants that play a role in surface water pollution may also be found in polluted groundwater, although their respective importance may differ.

Arsenic and Fluoride

Arsenic and fluoride have been recognized by the World Health Organization (WHO) as the most serious inorganic contaminants in drinking-water on a worldwide basis.

The metalloid arsenic can occur naturally in groundwater, as seen most frequently in Asia, including in China, India and Bangladesh. In the Ganges Plain of northern India and Bangladesh severe contamination of groundwater by naturally occurring arsenic affects 25% of water wells in the shallower of two regional aquifers.

Arsenic in groundwater can also be present where there are mining operations or mine waste dumps that will leach arsenic.

Natural fluoride in groundwater is of growing concern as deeper groundwater is being used, "with more than 200 million people at risk of drinking water with elevated concentrations." Fluoride can especially be released from acidic volcanic rocks and dispersed volcanic ash when water hardness is low. High levels of fluoride in groundwater is a serious problem in the Argentinean Pampas, Chile, Mexico, India, Pakistan, the East African Rift, and some volcanic islands (Tenerife).

In areas that have naturally occurring high levels of fluoride in groundwater which is used for drinking water, both dental and skeletal fluorosis can be prevalent and severe.

Pathogens

Waterborne diseases can be spread via a groundwater well which is contaminated with fecal pathogens from pit latrines.

Pathogens contained in feces can lead to groundwater pollution when they are given the opportunity to reach the groundwater, making it unsafe for drinking. Of the four pathogen types that are present in feces (bacteria, viruses, protozoa and helminths or helminth eggs), the first three can be commonly found in polluted groundwater, whereas the relatively large helminth eggs are usually filtered out by the soil matrix.

Groundwater that is contaminated with pathogens can lead to fatal fecal-oral transmission of diseases (e.g. cholera, diarrhoea).

Nitrate

Nitrate is the most common chemical contaminant in the world's groundwater and aquifers. In some low-income countries nitrate levels in groundwater are extremely high, causing significant health problems. It is also stable (it does not degrade) under high oxygen conditions.

Nitrate levels above 10 mg/L (10 ppm) in groundwater can cause "blue baby syndrome" (acquired methemoglobinemia). Drinking water quality standards in the European Union stipulate less than 50 mg/L for nitrate in drinking water.

However, the linkages between nitrates in drinking water and blue baby syndrome have been disputed in other studies. The syndrome outbreaks might be due to other factors than elevated nitrate concentrations in drinking water.

Elevated nitrate levels in groundwater can be caused by on-site sanitation, sewage sludge disposal and agricultural activities. It can therefore have an urban or agricultural origin.

Organic Compounds

Volatile organic compounds (VOCs) are a dangerous contaminant of groundwater. They are generally introduced to the environment through careless industrial practices. Many of these compounds were not known to be harmful until the late 1960s and it was some time before regular testing of groundwater identified these substances in drinking water sources.

Primary VOC pollutants found in groundwater include aromatic hydrocarbons such as BTEX compounds (benzene, toluene, ethylbenzene and xylenes), and chlorinated solvents including tetrachloroethylene (PCE), trichloroethylene (TCE), and vinyl chloride (VC). BTEX are important components of gasoline. PCE and TCE are industrial solvents historically used in dry cleaning processes and as a metal degreaser, respectively.

Other organic pollutants present in groundwater and derived from industrial operations are the polycyclic aromatic hydrocarbons (PAHs). Due to its molecular weight,

Naphthalene is the most soluble and mobile PAH found in groundwater, whereas benzo(a)pyrene is the most toxic one. PAHs are generally produced as byproducts by incomplete combustion of organic matter.

Organic pollutants can also be found in groundwater as insecticides and herbicides. As many other synthetic organic compounds, most pesticides have very complex molecular structures. This complexity determines the water solubility, adsorption capacity, and mobility of pesticides in the groundwater system. Thus, some types of pesticides are more mobile than others so they can more easily reach a drinking-water source.

Metals

Several trace metals can occurs naturally in some rocks or enter in the environmental from natural processes such as weathering. However, industrial activities such as mining and metallurgy, solid waste disposal, painting and enamel works can lead to severe groundwater pollution with elevated concentrations of toxic metals including lead, cadmium and chromium.

The migration of metals (and metalloids) in groundwater will be affected by several factors, in particular by chemical reactions which determine the partitioning of contaminants among different phases and species. Thus, the mobility of metals primarily depends on the pH and redox state of groundwater.

Others

Other organic pollutants include a range of organohalides and other chemical compounds, petroleum hydrocarbons, various chemical compounds found in personal hygiene and cosmetic products, drug pollution involving pharmaceutical drugs and their metabolites. Inorganic pollutants might include other nutrients such as ammonia and phosphate, and radionuclides such as uranium (U) or radon (Rn) naturally present in some geological formations. Saltwater intrusion is also an example of natural contamination, but is very often intensified by human activities.

Groundwater pollution is a worldwide issue. A study of the groundwater quality of the principal aquifers of the United States conducted between 1991 and 2004, showed that 23% of domestic wells had contaminants at levels greater than human-health benchmarks. Another study suggested that the major groundwater pollution problems in Africa, considering the order of importance are: (1) nitrate pollution, (2) pathogenic agents, (3) organic pollution, (4) salinization, and (5) acid mine drainage.

Causes

Naturally-occurring (Geogenic)

"Geogenic" refers to naturally occurring as a result from geological processes.

The natural arsenic pollution occurs because aquifer sediments contain organic matter that generates anaerobic conditions in the aquifer. These conditions result in the microbial dissolution of iron oxides in the sediment and, thus, the release of the arsenic, normally strongly bound to iron oxides, into the water. As a consequence, arsenic-rich groundwater is often iron-rich, although secondary processes often obscure the association of dissolved arsenic and dissolved iron.. Arsenic is found in groundwater most commonly as the reduced species arsenite and the oxidized species arsenate, being the acute toxicity of arsenite somewhat greater than that of arsenate. Investigations by WHO indicated that 20% of 25,000 boreholes tested in Bangladesh had arsenic concentrations exceeding 50 μg/l.

The occurrence of fluoride is close related to the abundance and solubility of fluoride-containing minerals such as fluorite (CaF_2). Considerably high concentrations of fluoride in groundwater are typically caused by a lack of calcium in the aquifer. Health problems associated with dental fluorosis may occur when fluoride concentrations in groundwater exceed 1.5 mg/l, which is the WHO guideline value since 1984.

The Swiss Federal Institute of Aquatic Science and Technology (EAWAG) has recently developed the interactive Groundwater Assessment Platform (GAP), where the geogenic risk of contamination in a given area can be estimated using geological, topographical and other environmental data without having to test samples from every single groundwater resource. This tool also allows the user to produce probability risk mapping for both arsenic and fluoride.

High concentrations of parameters like salinity, iron, manganese, uranium, radon and chromium, in groundwater, may also be of geogenic origin. This contaminants can be important locally but they are not as widespread as arsenic and fluoride.

On-site Sanitation Systems

A traditional housing compound near Herat, Afghanistan, where a shallow water supply well (foreground) is in close proximity to the pit latrine (behind the white greenhouse) leading to contamination of the groundwater.

Groundwater pollution with pathogens and nitrate can also occur from the liquids infiltrating into the ground from on-site sanitation systems such as pit latrines and septic tanks, depending on the population density and the hydrogeological conditions.

Factors controlling the fate and transport of pathogens are quite complex and the interaction among them is not well understood. If the local hydrogeological conditions (which can vary within a space of a few square kilometres) are ignored, simple on-site sanitation infrastructures such as pit latrines can cause significant public health risks via contaminated groundwater.

Liquids leach from the pit and pass the unsaturated soil zone (which is not completely filled with water). Subsequently, these liquids from the pit enter the groundwater where they may lead to groundwater pollution. This is a problem if a nearby water well is used to supply groundwater for drinking water purposes. During the passage in the soil, pathogens can die off or be adsorbed significantly, mostly depending on the travel time between the pit and the well. Most, but not all pathogens die within 50 days of travel through the subsurface.

The degree of pathogen removal strongly varies with soil type, aquifer type, distance and other environmental factors. For example, the unsaturated zone becomes "washed" during extended periods of heavy rain, providing hydraulic pathway for the quick pass of pathogens. It is difficult to estimate the safe distance between a pit latrine or a septic tank and a water source. In any case, such recommendations about the safe distance are mostly ignored by those building pit latrines. In addition, household plots are of a limited size and therefore pit latrines are often built much closer to groundwater wells than what can be regarded as safe. This results in groundwater pollution and household members falling sick when using this groundwater as a source of drinking water.

Sewage (Treated and Untreated)

Groundwater pollution can be caused by untreated waste discharge leading to diseases like skin lesions, bloody diarrhea and dermatitis. This is more common in locations having limited wastewater treatment infrastructure, or where there are systematic failures of the on-site sewage disposal system. Along with pathogens and nutrients, untreated sewage can also have an important load of heavy metals that may seep into the groundwater system.

The treated effluent from sewage treatment plants may also reach the aquifer if the effluent is infiltrated or discharged to local surface water bodies. Therefore, those substances that are not removed in conventional sewage treatment plants may reach the groundwater as well. For example, detected concentrations of pharmaceutical residues in groundwater were in the order of 50 ng/L in several locations in Germany. This is because in conventional sewage treatment plants, micro-pollutants such as hormones, pharmaceutical residues and other micro-pollutants contained in urine and feces are only partially removed and the remainder is discharged into surface water, from where it may also reach the groundwater.

Groundwater pollution can also occur from leaking sewers which has been observed for example in Germany. This can also lead to potential cross-contamination of drinking-water supplies.

Spreading wastewater or sewage sludge in agriculture may also be included as sources of faecal contamination in groundwater.

Fertilizers and Pesticides

Nitrate can also enter the groundwater via excessive use of fertilizers, including manure spreading. This is because only a fraction of the nitrogen-based fertilizers is converted to produce and other plant matter. The remainder accumulates in the soil or lost as run-off. High application rates of nitrogen-containing fertilizers combined with the high water-solubility of nitrate leads to increased runoff into surface water as well as leaching into groundwater, thereby causing groundwater pollution. The excessive use of nitrogen-containing fertilizers (be they synthetic or natural) is particularly damaging, as much of the nitrogen that is not taken up by plants is transformed into nitrate which is easily leached.

Poor management practices in manure spreading can introduce both pathogens and nutrients (nitrate) in the groundwater system.

The nutrients, especially nitrates, in fertilizers can cause problems for natural habitats and for human health if they are washed off soil into watercourses or leached through soil into groundwater. The heavy use of nitrogenous fertilizers in cropping systems is the largest contributor to anthropogenic nitrogen in groundwater worldwide.

Feedlots/animal corrals can also lead to the potential leaching of nitrogen and metals to groundwater. Over application of animal manure may also result in groundwater pollution with pharmaceutical residues derived from veterinary drugs.

The US Environmental Protection Agency (EPA) and the European Commission are seriously dealing with the nitrate problem related to agricultural development, as a major water supply problem that requires appropriate management and governance.

Runoff of pesticides may leach into groundwater causing human health problems from contaminated water wells. Pesticide concentrations found in groundwater are typically low, and often the regulatory human health-based limits exceeded are also very low. The organophosphorus insecticide monocrotophos (MCP) appears to be one of a few hazardous, persistent, soluble and mobile (it does not bind with minerals in soils) pesticides able to reach a drinking-water source. In general, more pesticide compounds are being detected as groundwater quality monitoring programs have become more extensive; however, much less monitoring has been conducted in developing countries due to the high analysis costs.

Commercial and Industrial Leaks

A wide variety of both inorganic and organic pollutants have been found in aquifers underlying commercial and industrial activities.

Ore mining and metal processing facilities are the primary responsible of the presence of metals in groundwater of anthropogenic origin, including arsenic. The low pH associated with acid mine drainage (AMD) contributes to the solubility of potential toxic metals that can eventually enter the groundwater system.

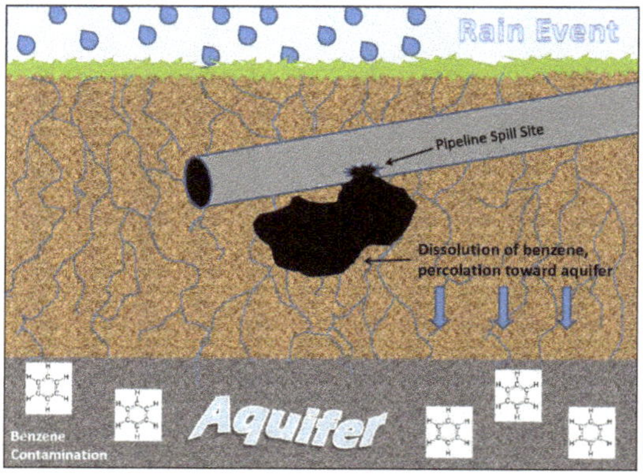

Oil spills associated with underground pipelines and tanks can release benzene and other soluble petroleum hydrocarbons that rapidly percolate down into the aquifer.

There is an increasing concern over the groundwater pollution by gasoline leaked from petroleum underground storage tanks (USTs) of gas stations. BTEX compounds are the most common additives of the gasoline. BTEX compounds, including benzene, have densities lower than water (1 g/ml). Similar to the oil spills on the sea, the non-miscible phase, referred to as Light Non-Aqueous Phase Liquid (LNAPL), will "float" upon the water table in the aquifer.

Chlorinated solvents are used in nearly any industrial practice where degreasing removers are required. PCE is a highly utilized solvent in the dry cleaning industry because of its cleaning effectiveness and relatively low cost. It has also been used for metal-degreasing operations. Because it is highly volatile, it is more frequently found in groundwater than in surface water. TCE has historically been used as a metal cleaning. The military facility Anniston Army Dept (ANAD) in the United States was placed on the US EPA Superfund National Priorities List (NPL) because of groundwater contamination with as much as 27 million pounds of TCE. Both PCE and TCE may degrade to vinyl chloride (VC), the most toxic chlorinated hydrocarbon.

Many types of solvents may have also been disposed illegally, leaking over time to the groundwater system.

Chlorinated solvents such as PCE and TCE have densities higher than water and the non-miscible phase is referred to as Dense Non-Aqueous Phase Liquids (DNAPL). Once they reach the aquifer, they will "sink" and eventually accumulate on the top of low-permeability layers.

Historically, wood-treating facilities have also release insecticides such as pentachlorophenol (PCP) and creosote into the environment, impacting the groundwater resources. PCP is a highly soluble and toxic obsolete pesticide recently listed in the Stockholm Convention on Persistent Organic Pollutants. PAHs and other semi-VOCs are the common contaminants associated with creosote.

Although non-miscible, both LNAPLs and DNAPLs still have the potential to slowly dissolve into the aqueous (miscible) phase to create a plume and thus become a long-term source of contamination. DNAPLs (chlorinated solvents, heavy PAHs, creosote, PCBs) tend to be difficult to manage as they can reside very deep in the groundwater system.

Hydraulic Fracturing

The recent growth of Hydraulic Fracturing ("Fracking") wells in the United States has raised concerns regarding its potential risks of contaminating groundwater resources. The Environmental Protection Agency (EPA), along with many other researchers, has been delegated to study the relationship between hydraulic fracturing and drinking water resources. While it is possible to perform hydraulic fracturing without having a relevant impact on groundwater resources if stringent controls und quality management measures are in place, there are a number of cases where groundwater pollution due to improper handling or technical failures was observed.

While the EPA has not found significant evidence of a widespread, systematic impact on drinking water by hydraulic fracturing, this may be due to insufficient systematic pre- and post- hydraulic fracturing data on drinking water quality, and the presence of other agents of contamination that preclude the link between shale oil/gas extraction and its impact.

Despite the EPA's lack of profound widespread evidence, other researchers have made significant observations of rising groundwater contamination in close proximity to major shale oil/gas drilling sites located in Marcellus*Ellsworth, William (2013). "Injection-Induced Earthquakes". Science AAAS.*</ref> (British Columbia, Canada). Within one kilometer of these specific sites, a subset of shallow drinking water consistently showed higher concentration levels of methane, ethane, and propane concentrations than normal. An evaluation of higher Helium and other noble gas concentration along with the rise of hydrocarbon levels supports the distinction between hydraulic fracturing fugitive gas and naturally occurring "background" hydrocarbon content. This contamination is speculated to be the result of leaky, failing, or improperly installed gas well casings.

Furthermore, it is theorized that contamination could also result from the capillary migration of deep residual hyper-saline water and hydraulic fracturing fluid, slowly flowing through faults and fractures until finally making contact with groundwater resources; however, many researchers argue that the permeability of rocks overlying shale formations are too low to allow this to ever happen sufficiently. To ultimately prove this theory, there would have to be traces of toxic trihalomethanes (THM) since they are often associated with the presence of stray gas contamination, and typically co-occur with high halogen concentrations in hyper-saline waters. Besides, highly saline waters are a common natural feature in deep groundwater systems.

While conclusions regarding groundwater pollution as the result to hydraulic fracturing fluid flow is restricted in both space and time, researchers have hypothesized that the potential for systematic stray gas contamination depends mainly on the integrity of the shale oil/gas well structure, along with its relative geological location to local fracture systems that could potentially provide flow paths for fugitive gas migration.

Though widespread, systematic contamination by hydraulic fracturing has been heavily disputed, one major source of contamination that has the most consensus among researchers of being the most problematic is site-specific accidental spillage of hydraulic fracturing fluid and produced water. So far, a significant majority of groundwater contamination events are derived from surface-level anthropogenic routes rather than the subsurface flow from underlying shale formations. Examples of such events include: a fracking fluid spillage in Acorn Fork Creek, Kentucky that caused a widespread death among aquatic species in 2007; a 420,000 gallon spillage of hyper-saline produced water that turned a once very-fertile farmland in New Mexico into a dead-zone in 2010; and a 42,000 gallon fracking fluid spillage in Arlington, Texas that necessitated an evacuation of over a 100 homes in 2015. While the damage can be obvious, and much more effort is being done to prevent these accidents from occurring so frequently, the lack of data from fracking oil spills continue to leave researchers in the dark. In many of these events, the data acquired from the leakage or spillage is often very vague, and thus would lead researchers to lacking conclusions.

Researchers from the Federal Institute for Geosciences and Natural Resources (BGR) conducted a modelling study for a deep shale-gas formation in the North German Basin. They concluded that the probability is small that the rise of fracking fluids through the geological underground to the surface will impact shallow groundwater.

Landfill Leachate

Leachate from sanitary landfills can lead to groundwater pollution.

Love Canal was one of the most widely known examples of groundwater pollution. In 1978, residents of the Love Canal neighborhood in upstate New York noticed high rates of cancer and an alarming number of birth defects. This was eventually traced to organic solvents and dioxins from an industrial landfill that the neighborhood had been built over and around, which had then infiltrated into the water supply and evaporated in basements to further contaminate the air. Eight hundred families were reimbursed for their homes and moved, after extensive legal battles and media coverage.

Other

Further causes of groundwater pollution are chemical spills from commercial or industrial operations, chemical spills occurring during transport (e.g. spillage of diesel fuels), illegal waste dumping, infiltration from urban runoff or mining operations, road salts, de-icing chemicals from airports and even atmospheric contaminants since groundwater is part of the hydrologic cycle.

The burial of corpses and their subsequent degradation may also pose a risk of pollution to groundwater.

Mechanisms

The passage of water through the subsurface can provide a reliable natural barrier to contamination but it only works under favorable conditions.

The stratigraphy of the area plays an important role in the transport of pollutants. An area can have layers of sandy soil, fractured bedrock, clay, or hardpan. Areas of karst topography on limestone bedrock are sometimes vulnerable to surface pollution from groundwater. Earthquake faults can also be entry routes for downward contaminant entry. Water table conditions are of great importance for drinking water supplies, agricultural irrigation, waste disposal (including nuclear waste), wildlife habitat, and other ecological issues.

Interactions with Surface Water

Although interrelated, surface water and groundwater have often been studied and managed as separate resources. Surface water seeps through the soil and becomes groundwater. Conversely, groundwater can also feed surface water sources. Sources of surface water pollution are generally grouped into two categories based on their origin.

Interactions between groundwater and surface water are complex. Consequently, groundwater pollution, sometimes referred to as groundwater contamination, is not as easily classified as surface water pollution. By its very nature, groundwater aquifers are susceptible to contamination from sources that may not directly affect surface water bodies, and the distinction of point vs. non-point source may be irrelevant. A spill or ongoing release of chemical or radionuclide contaminants into soil (located away from a surface water body) may not create point or non-point source pollution but can contaminate the aquifer below, creating a toxic plume.

Prevention

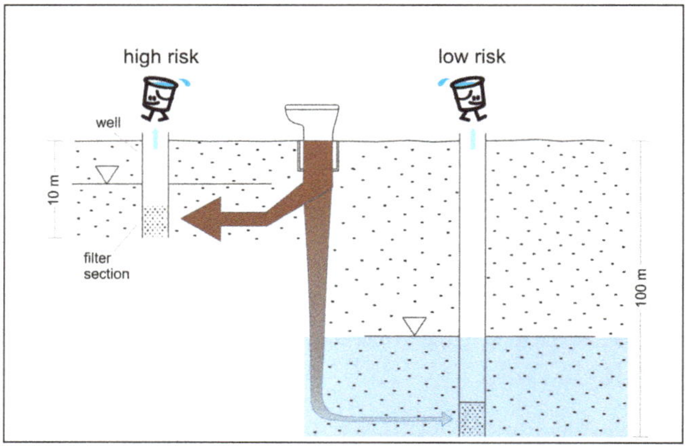

Schematic showing that there is a lower risk of groundwater pollution with greater depth of the water well.

Precautionary Principle

The precautionary principle, evolved from Principle 15 of the Rio Declaration on Environment and Development, is important in protecting groundwater resources from pollution. The precautionary principle provides that "*where there are threats of irreversible damage, lack of full scientific certainty shall not be used as reason for postponing cost-effective measures to prevent environmental degradation.*".

One of the six basic principles of the European Union (EU) water policy is the application of the precautionary principle.

Groundwater Quality Monitoring

Groundwater quality monitoring programs have been implemented regularly in many countries around the world. They are important components to understand the hydrogeological system, and for the development of conceptual models and aquifer vulnerability maps.

Groundwater quality must be regularly monitored across the aquifer to determine trends. Effective groundwater monitoring should be driven by a specific objective, for example, a specific contaminant of concern. Contaminant levels can be compared to the World Health Organization (WHO) guidelines for drinking-water quality. It is not rare that limits of contaminants are reduced as more medical experience is gained.

Sufficient investment should be given to continue monitoring over the long term. When a problem is found, action should be taken to correct it. Waterborne outbreaks in the United States decreased with the introduction of more stringent monitoring (and treatment) requirements in the early 90s.

The community can also help monitor the groundwater quality.

Land Zoning for Groundwater Protection

The development of land-use zoning maps has been implemented by several water authorities at different scales around the world. There are two types of zoning maps: aquifer vulnerability maps and source protection maps.

Aquifer Vulnerability Map

It refers to the intrinsic (or natural) vulnerability of a groundwater system to pollution. Intrinsically, some aquifers are more vulnerable to pollution than other aquifers. Shallow unconfined aquifers are more at risk of pollution because there are fewer layers to filter out contaminants.

The unsaturated zone can play an important role in retarding (and in some cases eliminating) pathogens and so must be considered when assessing aquifer vulnerability. The biological activity is greatest in the top soil layers where the attenuation of pathogens is generally most effective.

Preparation of the vulnerability maps typically involves overlaying several thematic maps of physical factors that have been selected to describe the aquifer vulnerability. The index-based parametric mapping method GOD developed by Foster and Hirata (1988) uses three generally available or readily estimated parameters, the degree of Groundwater hydraulic confinement, geological nature of the Overlying strata and Depth to groundwater. A further approach developed by the US EPA named DRASTIC employs seven hydrogeological factors to develop an index of vulnerability: Depth to water table, net Recharge, Aquifer media, Soil media, Topography (slope), Impact on the vadose zone, and hydraulic Conductivity.

There is a particular debate among hydrogeologist whether aquifer vulnerability should be established in a general (intrinsic) way for all contaminants, or specifically for each pollutant.

Source Protection Map

It refers to the capture areas around an individual groundwater source, such as a water well or a spring, to especially protect them from pollution. Thus, potential sources of degradable pollutants, such as pathogens, can be located at distances which travel times along the flowpaths are long enough for the pollutant to be eliminated through filtration or adsorption.

Analytical methods using equations to define groundwater flow and contaminant transport are the most widely used. The WHPA is a semi-analytical groundwater flow simulation program developed by the US EPA for delineating capture zones in a wellhead protection area.

The simplest form of zoning employs fixed-distance methods where activities are excluded within a uniformly applied specified distance around abstraction points.

Locating on-site Sanitation Systems

As the health effects of most toxic chemicals arise after prolonged exposure, risk to health from chemicals is generally lower than that from pathogens. Thus, the quality of the source protection measures is an important component in controlling whether pathogens may be present in the final drinking-water.

On-site sanitation systems can be designed in such a way that groundwater pollution from these sanitation systems is prevented from occurring. Detailed guidelines have been developed to estimate safe distances to protect groundwater sources from pollution from on-site sanitation. The following criteria have been proposed for safe siting (i.e. deciding on the location) of on-site sanitation systems:

- Horizontal distance between the drinking water source and the sanitation system.
    - o Guideline values for horizontal separation distances between on-site sanitation systems and water sources vary widely (e.g. 15 to 100 m horizontal distance between pit latrine and groundwater wells).
- Vertical distance between drinking water well and sanitation system.
- Aquifer type.
- Groundwater flow direction.
- Impermeable layers.
- Slope and surface drainage.
- Volume of leaking wastewater.
- Superposition, i.e. the need to consider a larger planning area.

As a very general guideline it is recommended that the bottom of the pit should be at least 2 m above groundwater level, and a minimum horizontal distance of 30 m between a pit and a water source is normally recommended to limit exposure to microbial contamination. However, no general statement should be made regarding the minimum lateral separation distances required to prevent contamination of a well from a pit latrine. For example, even 50 m lateral separation distance might not be sufficient in a strongly karstified system with a downgradient supply well or spring, while 10 m lateral separation distance is completely sufficient if there is a well developed clay cover layer and the annular space of the groundwater well is well sealed.

Legislation

Institutional and legal issues are critical in determining the success or failure of groundwater protection policies and strategies.

Sign near Mannheim, Germany indicating a zone as a dedicated "groundwater protection zone".

United States

In November 2006, the Environmental Protection Agency published the Ground Water Rule in the United States Federal Register. The EPA was worried that the ground water system would be vulnerable to contamination from fecal matter. The point of the rule was to keep microbial pathogens out of public water sources. The 2006 Ground Water Rule was an amendment of the 1996 Safe Drinking Water Act.

The ways to deal with groundwater pollution that has already occurred can be grouped into the following categories: containing the pollutants to prevent them from migrating further; removing the pollutants from the aquifer; remediating the aquifer by either immobilizing or detoxifying the contaminants while they are still in the aquifer (in-situ); treating the groundwater at its point of use; or abandoning the use of this aquifer's groundwater and finding an alternative source of water.

Management

Point-of-use Treatment

Portable water purification devices or "point-of-use" (POU) water treatment systems and field water disinfection techniques can be used to remove some forms of groundwater pollution prior to drinking, namely any fecal pollution. Many commercial portable water purification systems or chemical additives are available which can remove pathogens, chlorine, bad taste, odors, and heavy metals like lead and mercury.

Techniques include boiling, filtration, activated charcoal absorption, chemical disinfection, ultraviolet purification, ozone water disinfection, solar water disinfection, solar distillation, homemade water filters.

Arsenic removal filters (ARF) are dedicated technologies typically installed to remove arsenic. Many of these technologies require a capital investment and long-term maintenance. Filters in Bangladesh are usually abandoned by the users due to their high cost and complicated maintenance, which is also quite expensive.

Groundwater Remediation

Groundwater pollution is much more difficult to abate than surface pollution because groundwater can move great distances through unseen aquifers. Non-porous aquifers such as clays partially purify water of bacteria by simple filtration (adsorption and absorption), dilution, and, in some cases, chemical reactions and biological activity; however, in some cases, the pollutants merely transform to soil contaminants. Groundwater that moves through open fractures and caverns is not filtered and can be transported as easily as surface water. In fact, this can be aggravated by the human tendency to use natural sinkholes as dumps in areas of karst topography.

Pollutants and contaminants can be removed from ground water by applying various techniques thereby making it safe for use. Ground water treatment (or remediation) techniques span biological, chemical, and physical treatment technologies. Most ground water treatment techniques utilize a combination of technologies. Some of the biological treatment techniques include bioaugmentation, bioventing, biosparging, bioslurping, and phytoremediation. Some chemical treatment techniques include ozone and oxygen gas injection, chemical precipitation, membrane separation, ion exchange, carbon absorption, aqueous chemical oxidation, and surfactant enhanced recovery. Some chemical techniques may be implemented using nanomaterials. Physical treatment techniques include, but are not limited to, pump and treat, air sparging, and dual phase extraction.

Abandonment

If treatment or remediation of the polluted groundwater is deemed to be too difficult or expensive then abandoning the use of this aquifer's groundwater and finding an alternative source of water is the only other option.

Society and Culture

Examples

Hinkley, U.S.

The town of Hinkley, California (U.S.), had its groundwater contaminated with hexavalent chromium starting in 1952, resulting in a legal case against Pacific Gas & Electric (PG&E) and a multimillion-dollar settlement in 1996. The legal case was dramatized in the film *Erin Brockovich*, released in 2000.

California, U.S.

Nitrates and water contamination in California's Central Valley.

Walkerton, Canada

In the year 2000, groundwater pollution occurred in the small town of Walkerton, Canada leading to seven deaths in what is known as the Walkerton *E. Coli* outbreak. The water supply which was drawn from groundwater became contaminated with the highly dangerous O157:H7 strain of *E. coli* bacteria. This contamination was due to farm runoff into an adjacent water well that was vulnerable to groundwater pollution.

Lusaka, Zambia

The peri-urban areas of Lusaka, the capital of Zambia, have ground conditions which are strongly karstified and for this reason – together with the increasing population density in these peri-urban areas – pollution of water wells from pit latrines is a major public health threat there.

Concept of Groundwater flow

Consider the flow of ground water taking place within a small cube (of lengths Δx, Δy and Δz respectively the direction of the three areas which may also be called the elementary control volume) of a saturated aquifer as shown in Figure a.

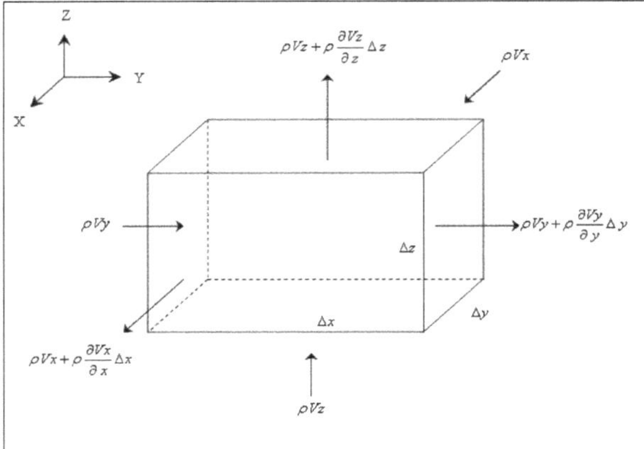

FIGURE a. Infinitesimal cube for deriving the equation of continuity of flow of ground water

It is assumed that the density of water (ρ) does not change in space along the three directions which implies that water is considered incompressible. The velocity components in the x, y and z directions have been denoted as v_x, v_y, v_z respectively.

Since water has been considered incompressible, the total incoming water in the cuboidal volume should be equal to that going out. Defining inflows and outflows as:

Inflows:

In x-direction: $\rho\ v_x\ (\Delta y.\Delta x)$

In y-direction: $\rho\ v_y\ (\Delta x.\Delta z)$

In z-direction: $\rho\ v_z\ (\Delta x.\Delta y)$

Outflows:

$$\text{In X-direction: } \rho\left(v_x+\frac{\partial vx}{\partial x}\Delta x\Delta x\right)(\Delta y.\Delta z)$$

$$\text{In Y-direction: } \rho\left(v_y+\frac{\partial vy}{\partial y}\Delta y\right)(\Delta x.\Delta z)$$

$$\text{In Z-direction: } \rho\left(v_z+\frac{\partial vz}{\partial z}\Delta z\right)(\Delta y.\Delta x)$$

The net mass flow per unit time through the cube works out to:

$$\left[\frac{\partial v_x}{\partial x}+\frac{\partial v_y}{\partial y}+\frac{\partial v_z}{\partial z}\right](\Delta x.\Delta y.\Delta z) \qquad (1)$$

Or

$$\frac{\partial v_x}{\partial x}+\frac{\partial v_y}{\partial y}+\frac{\partial v_z}{\partial z}=0 \qquad (2)$$

This is continuity equation for flow. But this water flow is due to a difference in potentiometric head per unit length in the direction of flow. A relation between the velocity and potentiometric gradient was first suggested by Henry Darcy, a French Engineer, in the mid nineteenth century. He found experimentally that the discharge 'Q' passing through a tube of cross sectional area 'A' filled with a porous material is proportional to the difference of the hydraulic head 'h' between the two end points and inversely proportional to the flow length 'L'.

It may be noted that the total energy (also called head, h) at any point in the ground water flow per unit weight is given as,

$$\text{h}=Z+\frac{p}{\gamma}+\frac{v^2}{2g} \qquad (3)$$

Where

Z is the elevation of the point above a chosen datum;

$\frac{p}{\gamma}$ is the pressure head, and

$\frac{v^2}{2g}$ is the velocity head

Since the ground water flow velocities are usually very small, $\frac{v^2}{2g}$ is neglected and $\text{h}=Z+\frac{p}{\gamma}$ is termed as the potentiometric head (or piezometric head in some texts),

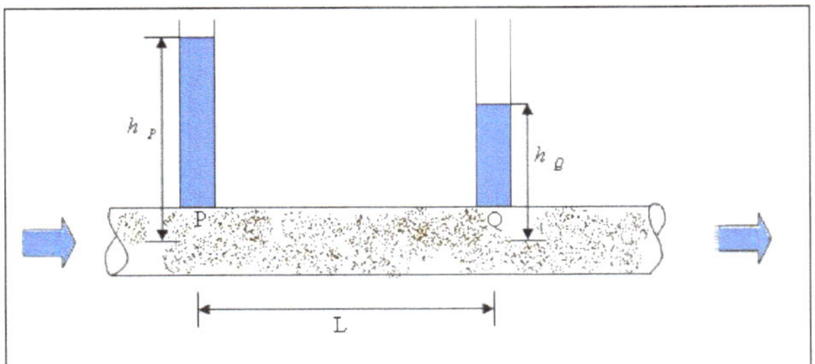

FIGURE b. Flow through a saturated porous medium

Thus,

$$\text{Q}\alpha\text{A}\bullet\left(\frac{\text{h}_\text{P}-\text{h}_\text{Q}}{\text{L}}\right) \qquad (4)$$

Introducing proportionality constant K, the expression becomes,

$$Q=K.A.\left(\frac{h_P - h_Q}{L}\right) \qquad (5)$$

Since the hydraulic head decreases in the direction of flow, a corresponding differential equation would be written as,

$$Q= -K.A.\left(\frac{dh}{dl}\right) \qquad (6)$$

Where (dh/dl) is known as hydraulic gradient.

The coefficient 'K' has dimensions of L/T, or velocity.

Thus the velocity of fluid flow would be:

$$v = \frac{Q}{A} = -K\left(\frac{dh}{dl}\right) \qquad (7)$$

It may be noted that this velocity is not quite the same as the velocity of water flowing through an open pipe. In an open pipe, the entire cross section of the pipe conveys water. On the other hand, if the pipe is filed with a porous material, say sand, then the water can only flow through the pores of the sand particles. Hence, the velocity obtained by the above expression is only an apparent velocity, with the actual velocity of the fluid particles through the voids of the porous material is many time more. But for our analysis of substituting the expression for velocity in the three directions x, y and z in the continuity relation, equation (2) and considering each velocity term to be proportional to the hydraulic gradient in the corresponding direction, one obtains the following relation,

$$\frac{\partial}{\partial x}\left(K_x \frac{\partial h}{\partial x}\right) + \frac{\partial}{\partial x}\left(K_y \frac{\partial h}{\partial y}\right) + \frac{\partial}{\partial z}\left(K_z \frac{\partial h}{\partial z}\right) = 0 \qquad (8)$$

Here, the hydraulic conductivities in the three directions (K_x, K_y and K_z) have been assumed to be different as for a general anisotropic medium. Considering isotropic medium with a constant hydraulic conductivity in all directions, the continuity equation simplifies to the following expression:

$$\frac{\partial^2 h}{\partial x^2} + \frac{\partial^2 h}{\partial y^2} + \frac{\partial^2 h}{\partial z^2} = 0 \qquad (9)$$

In the above equation, it is assumed that the hydraulic head is not changing with time, that is, a steady state is prevailing.

If now it is assumed that the potentiometric head changes with time at the location of the control volume, then there would be a corresponding change in the porosity of the aquifer even if the fluid density is assumed to be unchanged.

Important term:

Porosity: It is ratio of volume of voids to the total volume of the soil and is generally expressed as percentage.

Ground Water Flow Equations Under Unsteady State

For an unsteady case, the rate of mass flow in the elementary control volume is given by:

$$\rho\left[\frac{\partial v_x}{\partial x}+\frac{\partial v_y}{\partial y}+\frac{\partial v_z}{\partial z}\right]\Delta x\Delta y\Delta z=\frac{\partial M}{\partial t} \qquad (10)$$

This is caused by a change in the hydraulic head with time plus the porosity of the media increasing accommodating more water. Denoting porosity by the term 'n', a change in mass 'M' of water contained with respect to time is given by,

$$\frac{\partial M}{\partial t}=\frac{\partial}{\partial t}\left(\rho n\Delta x\Delta y\Delta z\right) \qquad (11)$$

Considering no lateral strain, that is, no change in the plan area Δx.Δy of the control volume, the above expression may be written as:

$$\frac{\partial M}{\partial t}=\frac{\partial \rho}{\partial t}\left(n\Delta x\Delta y\Delta z\right)+\frac{\partial}{\partial t}\left(n.\Delta z\right).\rho\Delta x\Delta y \qquad (12)$$

Where the density of water (ρ) is assumed to change with time. Its relation to a change in volume of the water V_w, contained within the void is given as:

$$\frac{d(V_w)}{V_w}=-\frac{d\rho}{\rho} \qquad (13)$$

The negative sign indicates that a reduction in volume would mean an increase in the density from the corresponding original values.

The compressibility of water, β, is defined as:

$$\beta=-\frac{\left[\dfrac{d(V_w)}{V_w}\right]}{dp} \qquad (14)$$

Where 'dp' is the change in the hydraulic head 'p' Thus,

$$\beta = \frac{d\rho}{\rho dp} \qquad (15)$$

That is,

$$d\rho = \rho\, dp\beta \qquad (16)$$

The compressibility of the soil matrix, α, is defined as the inverse of E_S, the elasticity of the soil matrix. Hence,

$$\frac{1}{\alpha} = E_S = -\frac{d(\sigma_z)}{\frac{d(\Delta z)}{\Delta z}} \qquad (17)$$

Where σ_z is the stress in the grains of the soil matrix.

Now, the pressure of the fluid in the voids, p, and the stress on the solid particles, σ_z must combine to support the total mass lying vertically above the elementary volume. Thus,

$$p+\sigma_z = \text{constant} \qquad (18)$$

Leading to,

$$d\sigma_z = -dp \qquad (19)$$

Thus,

$$\frac{1}{\alpha} = -\frac{dp}{\frac{d(\Delta z)}{\Delta z}} \qquad (20)$$

Also since the potentiometric head 'h' given by,

$$h = \frac{p}{\gamma} + Z \qquad (21)$$

Where Z is the elevation of the cube considered above a datum. We may therefore re-write the above as,

$$\frac{dh}{dz} = \frac{1}{\gamma}\frac{dp}{dz} + 1 \qquad (22)$$

First term for the change in mass 'M' of the water contained in the elementary volume, Equation 12, is

$$\frac{dp}{dt}.n.\Delta x\Delta y\Delta z \qquad (23)$$

This may be written, based on the derivations shown earlier, as equal to,

$$n.\rho.\beta.\frac{\partial p}{\partial t}.\Delta x\Delta y\Delta z \qquad (24)$$

Also the volume of soil grains, V_S, is given as,

$$V_S = (1-n)\Delta x\Delta y\Delta z \qquad (25)$$

Thus,

$$dV_S = [d(\Delta z)\text{-}d(n\Delta z)]\Delta x\Delta y \qquad (26)$$

Considering the compressibility of the soil grains to be nominal compared to that of the water or the change in the porosity, we may assume dVS to be equal to zero. Hence,

$$[d(\Delta z)\text{-}d(n\Delta z)]\Delta x\Delta y = 0 \qquad (27)$$

Or,

$$d(\Delta z) = d(n\Delta z) \qquad (28)$$

Which may substituted in second term of the expression for change in mass, M, of the elementary volume, changing it to,

$$\frac{\partial(n\Delta z)}{\partial t}\rho\Delta x\Delta y$$
$$= \frac{\partial(\Delta z)}{\partial t}\rho\Delta x\Delta y$$
$$= \rho\frac{\frac{\partial(\Delta z)}{\Delta z}}{\partial t}\Delta x\Delta y\Delta z$$
$$= \rho\alpha\frac{\partial p}{\partial t}\Delta x\Delta y\Delta z \qquad (29)$$

Thus, the equation for change of mass, M, of the elementary cubic volume becomes,

$$\frac{\partial M}{\partial t} = (\alpha + n\beta).\rho\frac{\partial p}{\partial t}\Delta x\Delta y\Delta z \qquad (30)$$

Combining Equation (30) with the continuity expression for mass within the volume, equation (10), gives the following relation:

$$-\rho\left[\frac{\partial v_x}{\partial x}+\frac{\partial v_y}{\partial y}+\frac{\partial v_z}{\partial z}\right]=(\alpha+n\beta)\rho\frac{\partial p}{\partial t} \qquad (31)$$

Assuming isotropic media, that is, $K_X=K_Y=K_Z=K$ and applying Darcy's law for the velocities in the three directions, the above equation simplifies to,

$$K\left[\frac{\partial^2 h}{\partial x^2}+\frac{\partial^2 h}{\partial y^2}+\frac{\partial^2 h}{\partial z^2}\right]=(\alpha+n\beta)\rho\frac{\partial p}{\partial t} \qquad (32)$$

Now, since the potentiometric (or hydraulic) head h is given as,

$$\mathrm{h}=\frac{p}{\gamma}+\mathrm{Z} \qquad (33)$$

The flow equation can be expressed as,

$$\left[\frac{\partial^2 h}{\partial x^2}+\frac{\partial^2 h}{\partial y^2}+\frac{\partial^2 h}{\partial z^2}\right]=(\alpha+n\beta)\frac{\gamma}{K}\frac{\partial h}{\partial t} \qquad (34)$$

The above equation is the general expression for the flow in three dimensions for an isotropic homogeneous porous medium. The expression was derived on the basis of an elementary control volume which may be a part of an unconfined or a confined aquifer.

Ground Water Flow Expressions for Ground Water Flow Unconfined And Confined Aquifers

Unsteady flow takes place in an unconfined and confined aquifer would be either due to:

- Change in hydraulic head (for unconfined aquifer) or potentiometric head (for confined aquifer) with time.
- And, or compressibility of the mineral grains of the soil matrix forming the aquifer.
- And, or compressibility of the water stored in the voids within the soil matrix.

We may visually express the above conditions as shown in Figure c, assuming an increase in the hydraulic (or potentiometric head) and a compression of soil matrix and pore water to accommodate more water.

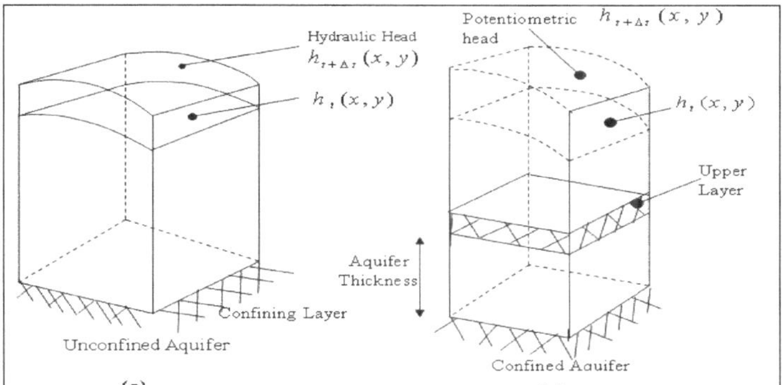

(a) (b)

FIGURE c. (a) Free surface of ground water table in unconfined aquifer
(b) Potentiometric surface in confined aquifer

Since storability S of a confined aquifer was defined as,

$$S = b\gamma(\alpha + n\beta) \qquad (35)$$

The flow equation for a confined aquifer would simplify to the following:

$$\left[\frac{\partial^2 h}{\partial x^2} + \frac{\partial^2 h}{\partial y^2} + \frac{\partial^2 h}{\partial z^2}\right] = \frac{S}{Kb}\frac{\partial h}{\partial t} \qquad (36)$$

Defining the transmissivity T of a confined aquifer as a product of the hydraulic conductivity K and the saturated thickness of the aquifer, b, which is:

$$T = K \bullet b \qquad (37)$$

The flow equation further reduces to the following for a confined aquifer,

$$\left[\frac{\partial^2 h}{\partial x^2} + \frac{\partial^2 h}{\partial y^2} + \frac{\partial^2 h}{\partial z^2}\right] = \frac{S}{T}\frac{\partial h}{\partial t} \qquad (38)$$

For unconfined aquifers, the storability S is given by the following expression,

$$S = S_y + h S_s \qquad (39)$$

Where S_y is the specific yield and S_s is the specific storage equal to $\gamma(\alpha + n\beta)$

Usually, S_s is much smaller in magnitude than S_y and may be neglected. Hence S under water table conditions for all practical purposes may be taken equal to S_y.

Two Dimensional Flow in Aquifers

Under many situations, the water table variation (for unconfined flow) in areal extent is not much, which means that there the ground water flow does not have much of a vertical velocity component. Hence, a two – dimensional flow situation may be approximated for these cases. On the other hand, where there is a large variation in the water table under certain situation, a three dimensional velocity field would be the correct representation as there would be significant component of flow in the vertical direction apart from that in the horizontal directions. This difference is shown in the illustrations given in Figure.

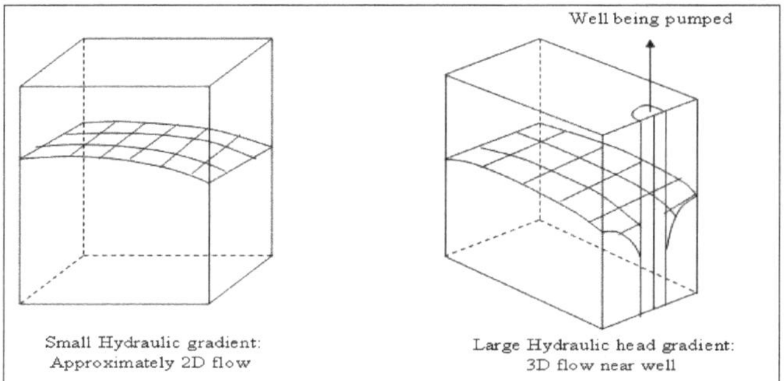

FIGURE d. Difference between small and large hydraulic gradients of ground water table

In case of two dimensional flow, the equation flow for both unconfined and confined aquifers may be written as,

$$\left[\frac{\partial^2 h}{\partial x^2}+\frac{\partial^2 h}{\partial y^2}\right]=\frac{S}{T}\frac{\partial h}{\partial t} \tag{40}$$

There is one point to be noted for unconfined aquifers for hydraulic head (or water table) variations with time. It is that the change in the saturated thickness of the aquifer with time also changes the transmissivity, T, which is a product of hydraulic conductivity K and the saturated thickness h. The general form of the flow equation for two dimensional unconfined flow is known as the Boussinesq equation and is given as,

$$\frac{\partial}{\partial x}\left(h\frac{\partial h}{\partial x}\right)+\frac{\partial}{\partial y}\left(h\frac{\partial h}{\partial y}\right)=\frac{S_y}{K}\frac{\partial h}{\partial t} \tag{41}$$

Where S_y is the specific yield.

If the drawdown in the unconfined aquifer is very small compared to the saturated thickness, the variable thickness of the saturated zone, h, can be replaced by an average thickness, b, which is assumed to be constant over the aquifer.

For confined aquifer under an unsteady condition though the potentiometric surface declines, the saturated thickness of the aquifer remains constant with time and is equal to an average value 'b'. Solving the ground water flow equations for flow in aquifers require the help of numerical techniques, except for very simple cases.

Two Dimensional Seepage Flow

Another example of two dimensional flow would that be when the flow can be approximated. to be taking place in the vertical plane. Such situations might occur as for the seepage taking place below a dam as shown in Figure.

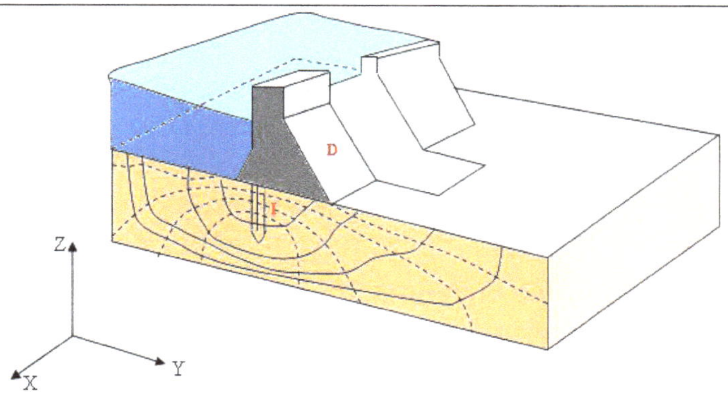

FIGURE e. Seepage flow below a concrete gravity dam

Under steady state conditions, the general equation of flow, considering an isotropic porous medium would be,

$$\frac{\partial^2 h}{\partial y^2} + \frac{\partial^2 h}{\partial z^2} = 0 \qquad (42)$$

However, solving the above Equation (42) for would require advanced analytical methods or numerical techniques. More about seepage flow would be discussed in the later session.

Steady One Dimensional Flow in Aquifers

Some simplified cases of ground water flow, usually in the vertical plane, can be approximated by one dimensional equation which can then be solved analytically.

Confined Aquifers

If there is a steady movement of ground water in a confined aquifer, there will be a gradient or slope to the potentiometric surface of the aquifer. The gradient, again, would be decreasing in the direction of flow. For flow of this type, Darcy's law may be used directly.

Aquifer with Constant Thickness

This situation may be shown as in Figure f.

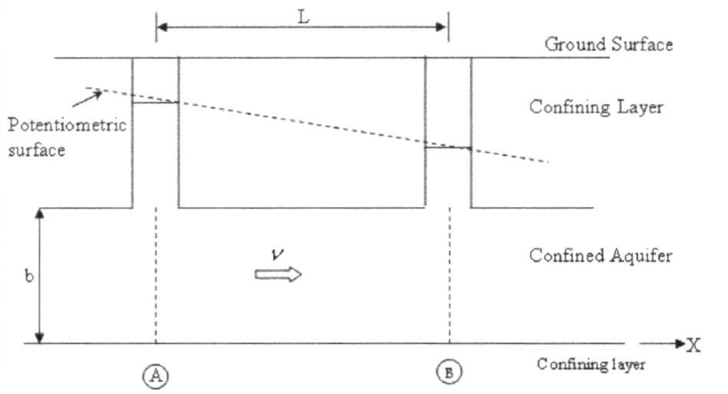

FIGURE F. Flow through an aquifer of constant thckness

Assuming unit thickness in the direction perpendicular to the plane of the paper, the flow rate 'q' (per unit width) would be expressed for an aquifer of thickness 'b',

$$q=b*1*v \qquad (43)$$

According to Darcy's law, the velocity 'v' is given by,

$$v=-K\frac{\partial h}{\partial x} \qquad (44)$$

Where h, the potentiometric head, is measured above a convenient datum. Note that the actual value of 'h' is not required, but only its gradient $\frac{\partial h}{\partial x}$ in the direction of flow, x, is what matters. Here is K is the hydraulic conductivity

Hence,

$$q=bK\frac{\partial h}{\partial x} \qquad (45)$$

The partial derivative of 'h' with respect to 'x' may be written as normal derivative since we are assuming no variation of 'h' in the direction normal to the paper. Thus

$$q=-bK\frac{dh}{dx} \qquad (46)$$

For steady flow, q should not vary with time, t, or spatial coordinate, x. hence,

$$\frac{dq}{dx}=-bK\frac{d^2h}{dx^2}=0 \qquad (47)$$

Since the width, b, and hydraulic conductivity, K, of the aquifer are assumed to be constants, the above equation simplifies to:

$$\frac{d^2h}{dx^2} = 0 \qquad (48)$$

Which may be analytically solved as,

$$h = C_1x+C_2 \qquad (49)$$

Selecting the origin of coordinate x at the location of well A (as shown in Figure f), and having a hydraulic head, h_A and also assuming a hydraulic head of well B, located at a distance L from well A in the x-direction and having a hydraulic head h_B, we have:

$$h_A = C_1.0+C_2 \text{ and}$$
$$h_B = C_1.L+C_2$$

Giving

$$C_1=h\text{-}h_A/L \text{ and } C_2=h_A \qquad (50)$$

Thus the analytical solution for the hydraulic head 'h' becomes:

$$H = \frac{h_B - h_A}{L}x + h_A \qquad (51)$$

Aquifer with Variable Thickness

Consider a situation of one- dimensional flow in a confined aquifer whose thickness, b, varies in the direction of flow, x, in a linear fashion as shown in Figure g.

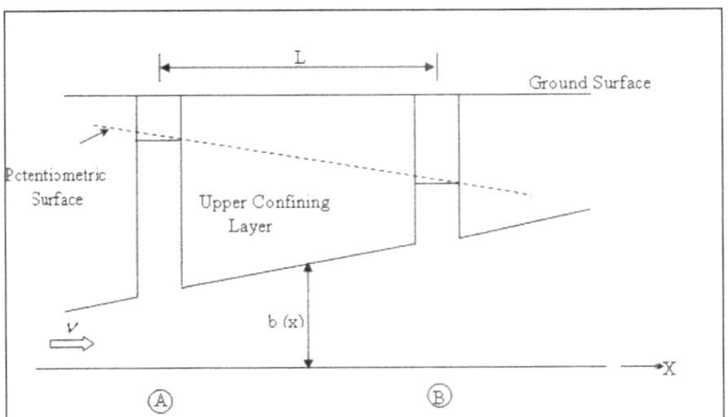

FIGURE g. Flow through an aquifer with variable thickness

The unit discharge, q, is now given as,

$$q = -b(x)K\frac{dh}{dx} \qquad (52)$$

Where K is the hydraulic conductivity and dh/dx is the gradient of the potentiometric surface in the direction of flow, x.

For steady flow, we have,

$$\frac{dq}{dx} = -K\left[\frac{db}{dx}\frac{dh}{dx} + b\frac{d^2h}{dx^2}\right] = 0 \qquad (53)$$

Which may be simplified, denoting $\frac{db}{dx}$ as b'

$$b\frac{d^2h}{dx^2} + b'\frac{dh}{dx} = 0 \qquad (54)$$

A solution of the above differential equation may be found out which may be substituted for known values of the potentiometric heads h_A and h_B in the two observation wells A and B respectively in order to find out the constants of integration.

Unconfined Aquifers

In an unconfined aquifer, the saturated flow thickness, h is the same as the hydraulic head at any location, as seen from Figure h:

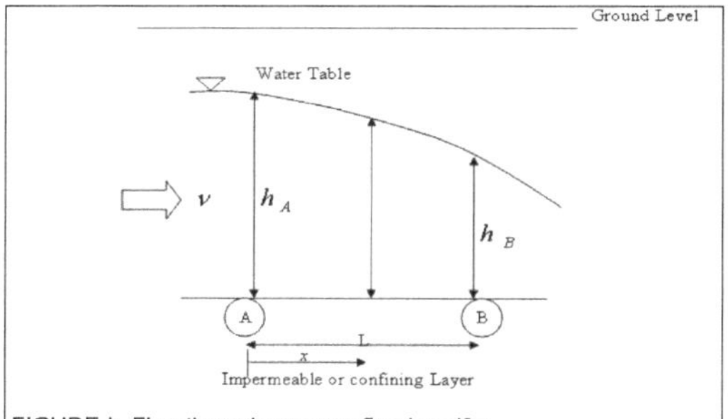

FIGURE h. Flow through an unconfined aquifer

Considering no recharge of water from top, the flow takes place in the direction of fall of the hydraulic head, h, which is a function of the coordinate, x taken in the flow direction. The flow velocity, v, would be lesser at location A and higher at B since the saturated flow thickness decreases. Hence v is also a function of x and increases in the direction of flow. Since, v, according to Darcy's law is shown to be,

$$v = K\frac{dh}{dx} \qquad (55)$$

the gradient of potentiometric surface, dh/dx, would (in proportion to the velocities) be smaller at location A and steeper at location B. Hence the gradient of water table in unconfined flow is not constant, it increases in the direction of flow.

This problem was solved by J.Dupuit, a French hydraulician, and published in 1863 and his assumptions for a flow in an unconfined aquifer is used to approximate the flow situation called Dupuit flow. The assumptions made by Dupuit are:

- The hydraulic gradient is equal to the slope of the water table, and
- For small water table gradients, the flow-lines are horizontal and the equi-potential lines are vertical.

The second assumption is illustrated in Figure.

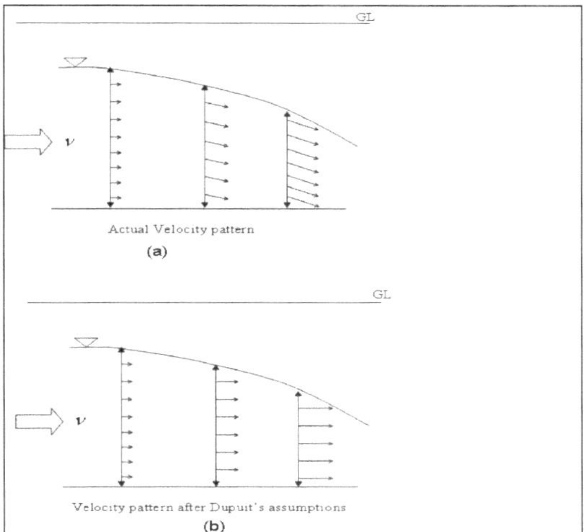

FIGURE i. (a) Actual velocity pattern in ground water flow and (b) Assumption of Dupuit regarding ground water flow

Solutions based on the Dupuit's assumptions have proved to be very useful in many practical purposes. However, the Dupuit assumption do not allow for a seepage face above an outflow side.

An analytical solution to the flow would be obtained by using the Darcy equation to express the velocity, v, at any point, x, with a corresponding hydraulic gradient $\frac{dh}{dx}$ as,

$$v = -K\frac{dh}{dx} \qquad (56)$$

Thus, the unit discharge, q, is calculated to be,

$$q = -Kh\frac{dh}{dx} \qquad (57)$$

Considering the origin of the coordinate x at location A where the hydraulic head us h_A and knowing the hydraulic head h_B at a location B, situated at a distance L from A, we may integrate the above differential equation as,

$$\int_0^L q\,dx = -K\int_{h_A}^{h_B} h\,dh \tag{58}$$

Which, on integration, leads to,

$$q\,x\Big|_0^L = -K.\frac{h^2}{2}\Bigg|_{h_A}^{h_B} \tag{59}$$

Or,

$$q.L = K\left[\frac{h_B^2}{2} - \frac{h_A^2}{2}\right] \tag{60}$$

Rearrangement of above terms leads to, what is known as the Dupuit equation:

$$q = -\frac{1}{2}K\left[\frac{h_B^2 - h_A^2}{L}\right] \tag{61}$$

An example of the application of the above equation may be for the ground water flow in a strip of land located between two water bodies with different water surface elevations, as in Figure j.

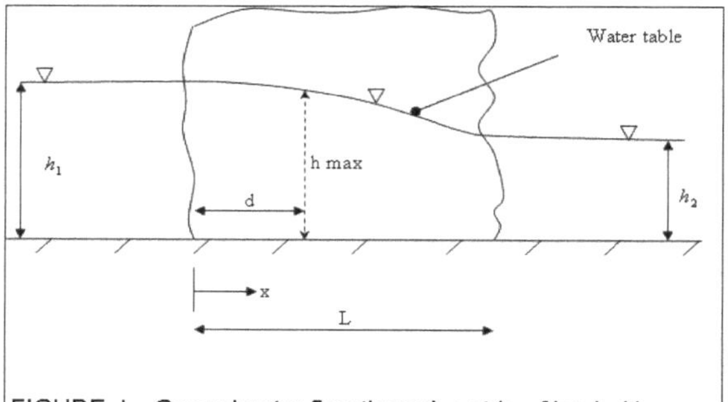

FIGURE j . Ground water flow through a strip of land with difference in water surface elevation on either side.

The equation for the water table, also called the phreatic surface may be derived from Equation (61) as follows:

$$h = \sqrt{h_1^2 - \left(h_1^2 - h_2^2\right)\frac{x}{L}} \tag{62}$$

In case of recharge due to a constant infiltration of water from above the water table rises to a many as shown in Figure k.

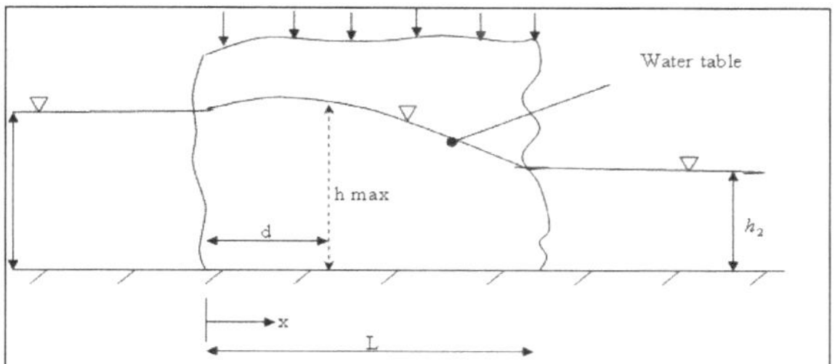

FIGURE k. Ground water flow through a strip of land with infiltration from above

There is a difference with the earlier cases, as the flow per unit width, q, would be increasing in the direction of flow due to addition of water from above. The flow may be analysed by considering a small portion of flow domain as shown in Figure l.

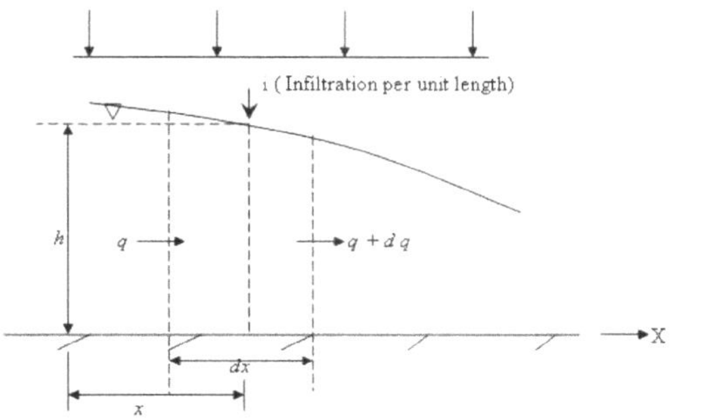

FIGURE l. Definition of terms for flow analysis for the case shown in Figure k

Considering the infiltration of water from above at a rate i per unit length in the direction of ground water flow, the change in unit discharge d_q is seen to be,

$$d_q = i.dx \qquad (63)$$

Or,

$$\frac{dq}{dx} = i \qquad (64)$$

From Darcy's law,

$$q = -K.h.\frac{dh}{dx} = -\frac{1}{2}\ k\frac{d\left(h^2\right)}{dx^2} \qquad (65)$$

$$\frac{dq}{dx} = -\frac{1}{2}\ K\frac{d^2\left(h^2\right)}{dx^2} \qquad (66)$$

Substituting the expression for $\frac{dq}{dx}$ we have,

$$-\frac{1}{2}\ K\frac{d^2\left(h^2\right)}{dx^2}=i \tag{67}$$

Or,

$$\frac{d^2\left(h^2\right)}{dx^2}=\frac{2.i}{k} \tag{68}$$

The solution for this equation is of the form,

$$Kh^2+2x^2=C_1x+C_2 \tag{69}$$

If, now, the boundary condition is applied as,

At x = o, h = h_1, and

At x = L, h = h_2

The equation for the water table would be:

$$h=\sqrt{h_1^2-\left(\frac{h_1^2-h_1^2}{L}\right)x+\frac{i}{K}(L-x)x} \tag{70}$$

And,

$$q_x=q_0+2x \tag{71}$$

Where q_0 is the unit discharge at the left boundary, x = o, and may be found out to be,

$$q_0=\left(\frac{h_1^2-h_1^2}{2L}\right)-\frac{iL}{2} \tag{72}$$

Which gives an expression for unit discharge q_x at any point x from the origin as,

$$q_x=K\left(\frac{h_1^2-h_1^2}{2L}\right)-i\left(\frac{L}{2}-x\right) \tag{73}$$

For no recharge due to infiltration, i = o and the expression for q_x is then seen to become independent of x, hence constant, which is expected.

Flow Dynamics

It is common for water resources engineers to design a water system involving flow of water from one place to another, usually passing a variety of structures on the way some of them meant for controlling the flow quantity. Rivers and artificial channels, like canals, convey water with a free surface, that is, the surface of water being exposed to air as opposed to flow of water in pipes. It is easy to visualize that for any such open channel flow, as they are called; the presence or absence of a hydraulic structure controls the position of the free surface of water. Knowing the mathematical description of flowing water, it is possible to compute the water surface profile, which is important for example in designing the height of the channel walls of the water conveying system.

Another example, the case of river flow obstruction by the presence dam may be mentioned. The water level of the river increases on construction of the dam and it is essential to know the maximum possible rise, perhaps during the maximum flood, in order to know the degree of submergence of the land behind the dam. Barrages are low height structures, and hence, the rise of water will not be occurring uniformly across the river, again due to the difference of gate operation.

In this lesson, the behavior and corresponding mathematical description of flow in open channels are reviewed in order to utilize them in designing water resources systems.'

Flow in Natural Rivers

Figure a below shows a river carrying a low discharge.

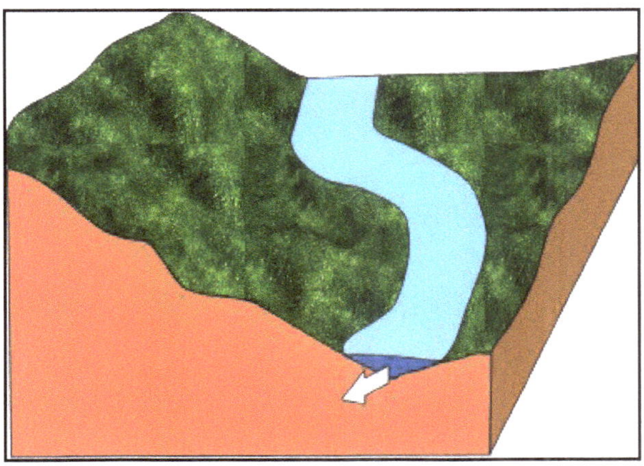

A river flowing within its banks .

When the water surface of the river just touches its banks, the discharge flowing through the river at this stage is called the "bank full discharge". It is also sometimes called the "dominant discharge". If the discharge in the river increases, the water will overflow the banks and would spill over to the adjacent land, called the flood plains (Figure b).

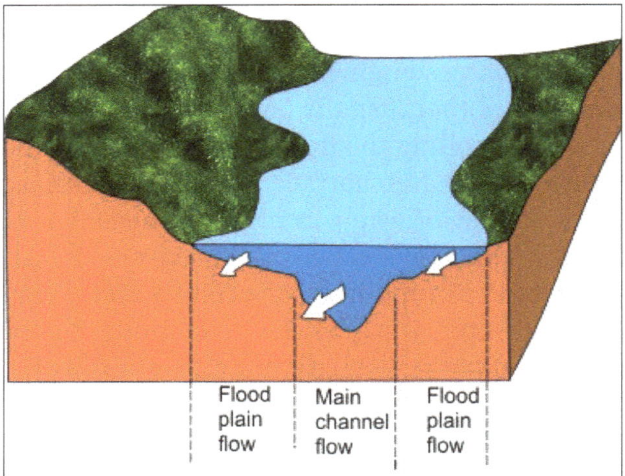

FIGURE .b A river flowing its banks during flood

Though the amount of discharge flowing through the river is of interest to the water resources engineer it cannot be measured directly by any instruments. Rather, an indirect method is used which requires knowledge of the velocity distribution in a river or an open channel.

If we plot the velocity profile across a river, as shown in Figure a, it would actually vary in three dimensions. Figure c shows the variation of velocity at the water surface.

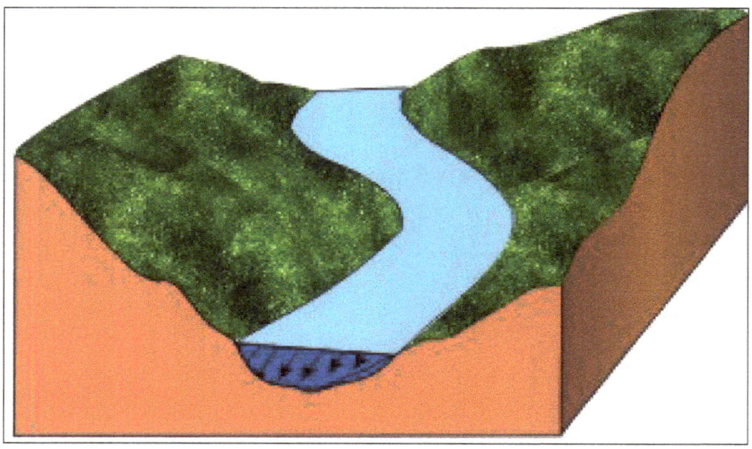

FIGURE C. Variation of surface velocity across a river section.

It may be observed that velocity is highest at the center of the river but is zero at the banks. If a velocity profile were plotted on another horizontal plane at certain depth of the river, there too the velocity profile would be found to be similar in shape, but smaller in magnitude (Figure d).

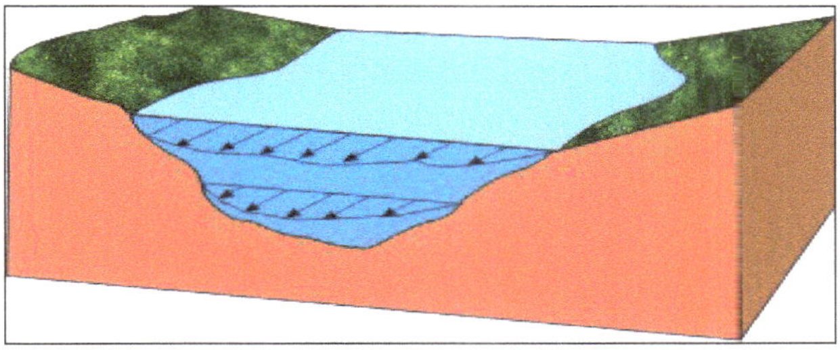

FIGURE d. Velocity variation of a river section at two levels.

Similarly the velocity profile of the river flowing in flood would be as shown in Figure e, showing that the velocities over the flood plains is smaller compared to the main stream flow.

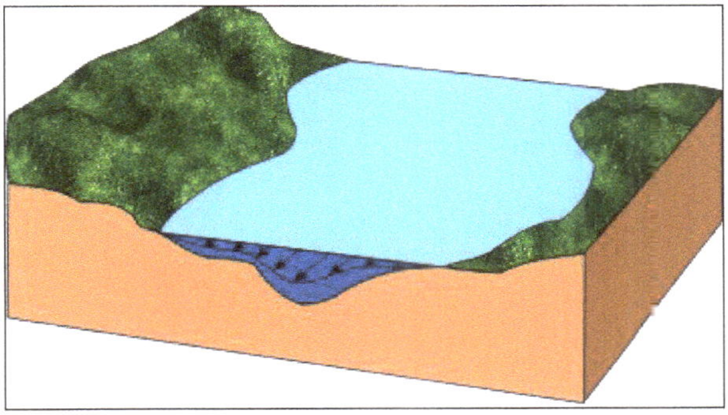

FIGURE e. Surface velocity profile across a river section for a river flowing in flood.

If we now take a look at the variation of velocity in a vertical plane within a river, and we plot them along different vertical lines across the river, then we may find the velocity profiles similar to those shown in Figure f.

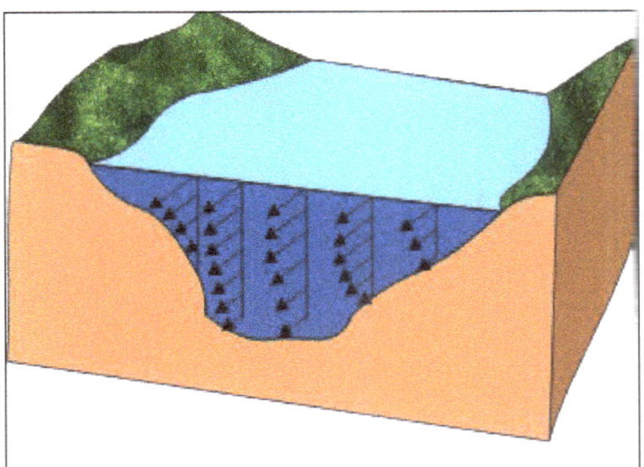

FIGURE f. Vertical velocity profiles in a river section

In order to measure the discharge being conveyed in a river, the velocity profile or the average velocity at a number of equally spaced sections are measured, as in Figure f. The total discharge is then approximately taken equal to the sum of the discharges passing through each segment.

Another way of depicting the velocity variation across a river cross-section is to plot "Isovels", which are actually the locus of points having equal velocity (Figure g).

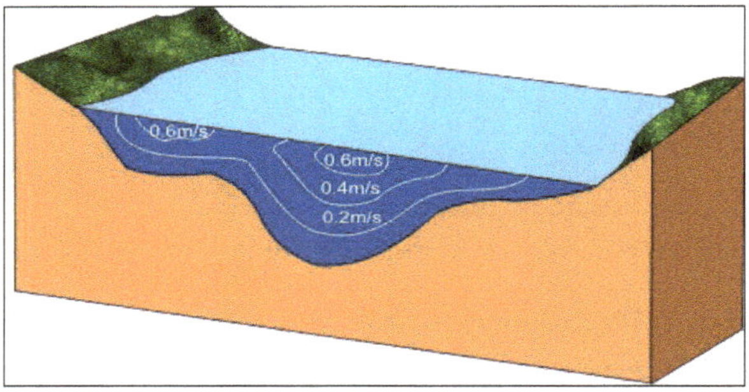

FIGURE g. Isovels showing contours of typical equal velocities across a typical river section.

It has been observed through experiments that a plot of velocity in the vertical plane would show that the maximum velocity occurs slightly below the surface (Figure h) for a typical river flow.

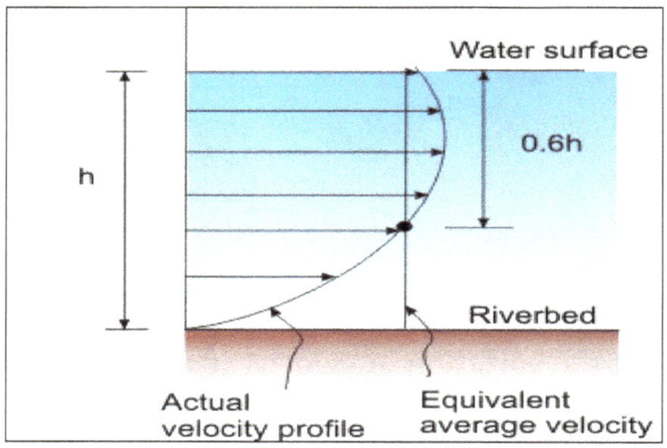

FIGURE h. Vertical velocity distribution and average velocity for flow in a river.

It has further been observed that an equivalent average velocity is almost equal to the actual velocity measured at 0.6 depth.

Variation of Discharge with River Stage

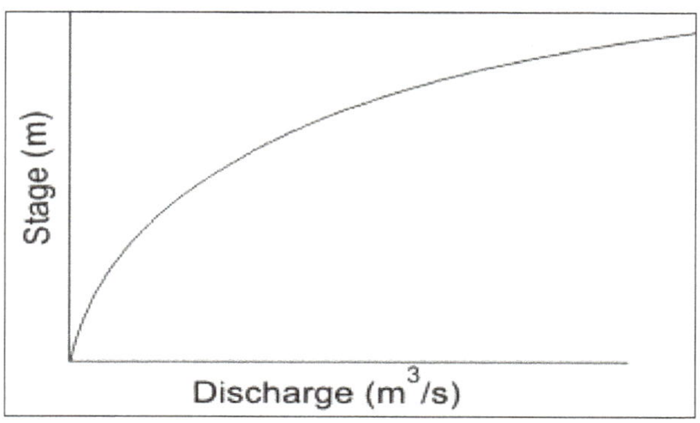

FIGURE i. Stage-discharge curve for a river.

The water level in a river is sometimes called the "stage" and as this varies, there is a proportional change in the total discharge conveyed. For each point of a river, the relation between stage and discharge is unique but a general form is found to be as shown in Figure i.

The general mathematical description for the stage-discharge relation is given as:

$$Q = k\left(h - h_0\right)^m$$

Where h is the gauge corresponding to a discharge Q and ho is the corresponding to zero discharge k and m are constants. If the variables (Q and H) are plotted on a log-log graph, then it generally plots in a straight line as:

$$\log Q = m \log\left(h - h_0\right) + \log k$$

Flow Variation Along River Length

It may be interpreted from Figures d or f that the velocity in a river cross section actually varies from bank to bank and from riverbed to free water surface and hence, can be called a two dimensional variation in a vertical plane. However, for engineering purposes it is, sufficient, generally, to use an equivalent velocity in the direction of river motion (perpendicular to river cross section) which may be obtained by dividing the total discharge by the cross sectional area. In a natural river, therefore, these flow velocities may vary from section to section (Figure j).

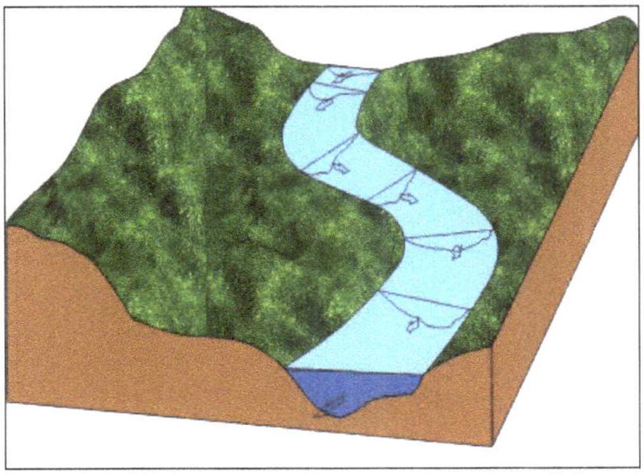

FIGURE j. Variation of cross section and velocity along the length of a river showing vertical section at each point.

If we now consider an axis along the length of the river, the total energy (H) is given as:

$$H = Z + h + \frac{V^2}{2g}$$

We may plot the total energy as shown in Figure k, where the variables are as follows:

- Z: Height of riverbed above a datum
- h: Depth of water
- V: Average velocity at a section
- $\frac{V^2}{2g}$: Kinetic energy head

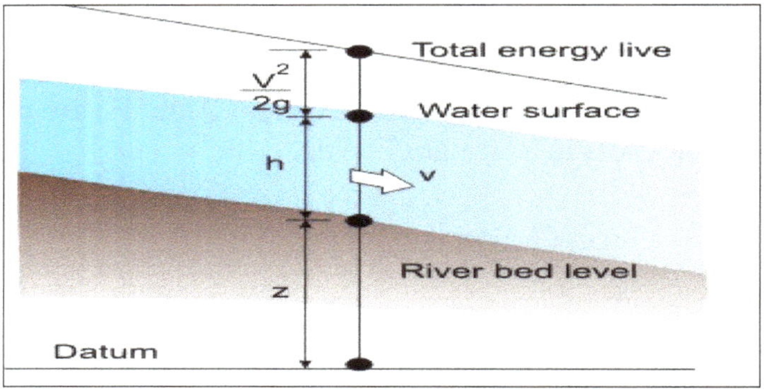

FIGURE k. Variation of total energy along a river.

Since the cross section, bed slope and flow resistance vary along a river length, the depth and velocity would vary correspondingly. However, if a short stretch of a river section is taken, then the variations in riverbed, water surface and the total energy may be considered as linear (Figure l).

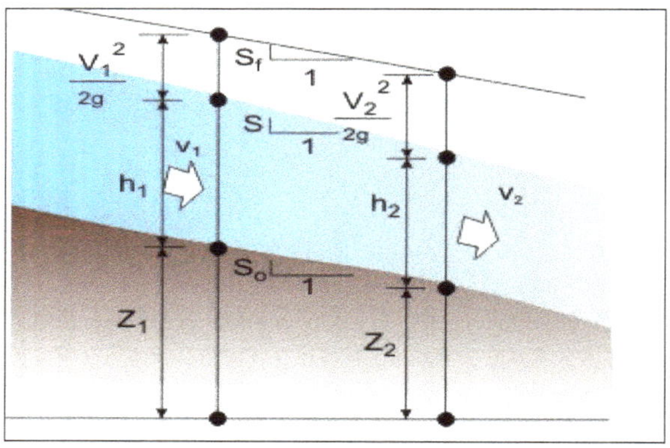

FIGURE l. Short section between two points along a river.

In Figure l, three slopes have been marked, which are:

- S_o: Riverbed slope
- S: Water surface slope
- S_f: Energy surface slope

Since the total energy of flowing water reduces along the river length due to friction the "energy surface slope" is generally termed as the "friction slope". The energy loss in a river or an open channel occurs mostly due to the resistance at the channel sides, as the turbulent characteristics of the flowing water implies a smaller loss internally within the water body itself.

It has been nearly 200 years when scientists first attempted to mathematically express (or "model") the friction slope in terms of known variables like average velocity, cross section properties and riverbed slope. One of the earliest models for friction slope S_f or, in effect, the channel resistance was derived from the considerations of "uniform flow" (Figure m) where the flow variables and cross section are supposed to remain constant over a short reach.

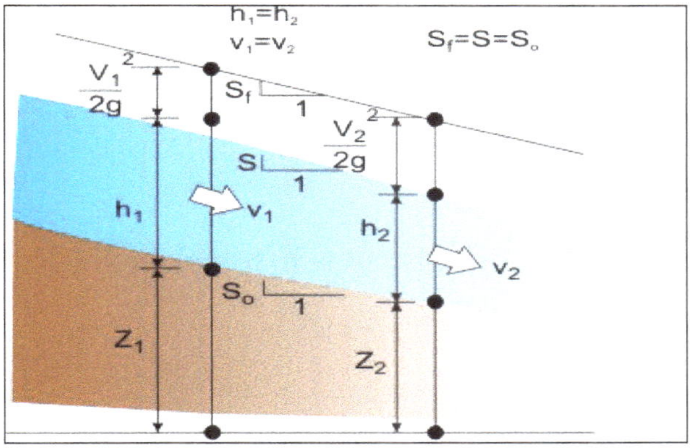

FIGURE m. Uniform flow along a channel reach.

If we take small volume of fluid from these two sections we may make a free body diagram of the forces acting on it (Figure n).

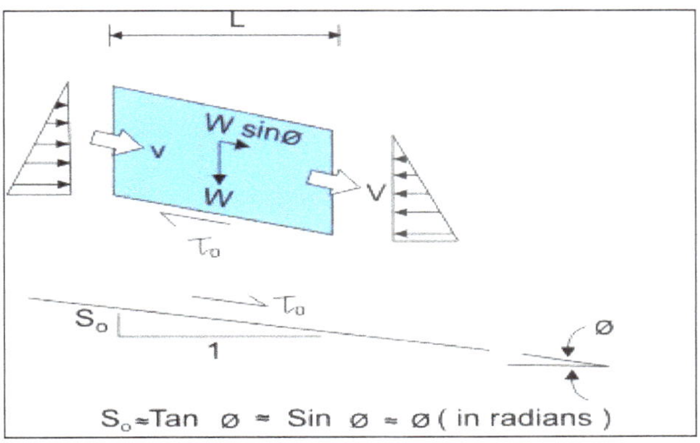

FIGURE n. Free body diagram of forces on a control volume in uniform flow.

The variables represented in the figure are as follows

- W: Weight of water contained in the control volume.
- V: Inflow velocity, which is the same as the outflow velocities.

- θ: Angle of slope river bed, which is also equal to that water surface and friction slopes.
- τ_0 : Shear stress due to friction acting on the control volume of fluid from the river bed and all along the periphery, though in Figure n only the resistance due to the riverbed is shown.

Equating the forces and noting that the inflowing and out flowing momenta are equal as well as the pressure forces at either end of the control volume one obtains:

$$\tau_0 PL = W\sin\theta = pgAL\sin\theta$$

Where the remaining variables are:

- P: wetted perimeter.
- A: Cross section of flow area.
- L: Length of control volume.

Assuming θ to be very small and nearly equal to bed slope, we have,

$$\tau_0 = pgRS_0$$

Assuming a state of rough turbulent flow, as is the case for natural rivers and channels, one may write,

$$\tau_0 \alpha V^2 \qquad \text{or} \qquad \tau_0 = kV^2$$

Substituting into equation above,

$$V = \sqrt{\frac{\rho g}{k} RS_0}$$

This may be written as,

$$V = C\sqrt{RS}$$

This is known as Chezy equation after the French hydraulic engineer. Antoine Chezy who first proposed the formula around 1768 while designing a canal for Paris water supply. The constant C in equation (8) actually varies depending on Reynolds number and boundary roughness.

In 1869, Swiss engineers, Ganguillet and Kutter proposed an elaborate formula for Chezy's C which they derived from actual discharge data from the river Mississippi and a wide range of natural and artificial channels in Europe. The formula, in metric units, is given as,

$$C = 0.552\left(\frac{41.6 + \frac{1.811}{n} + \frac{0.00281}{S_0}}{1 + \left[41.65 + \left(\frac{0.00281}{S_0}\right)\right]\frac{n}{\sqrt{R}}}\right)$$

Where n is a coefficient known as Kutter's n, and is dependent solely on the boundary roughness.

In 1889, Robert Manning's, an Irish engineer proposed another formula for the evaluation of the Chezy coefficient, which was later simplified to:

$$C = \frac{R^{1/6}}{n}$$

From Equation earlier, the Manning equation may be written as:

$$V = \frac{1}{n}R^{2/3}S_0^{\ 1/2}$$

Where the Manning n is numerically equivalent to Kutter's n.

Uniform Flow in Channels of Simple Cross Section

For problems concerning the steady uniform flow in rivers and open channels, the Manning's equation is commonly used in India. The depth of water corresponding to a discharge in a channel or river under uniform flow conditions is called "normal depth". By combining the continuity equation with that of Mannings, one obtains,

$$Q = \frac{1}{n}AR^{2/3}S^{1/2}$$

Where the variables have been defined in the earlier sections

One may also write equation as follows,

$$Q = K\sqrt{S}$$

Where $K = \frac{AR^{2/3}}{n}$ also called Conveyance, is often necessary to find out the normal depth of flow corresponding to a discharge Q, flowing in a channel for which equation may be rearranged as,

$$AR^{2/3} = \frac{nQ}{S^{1/2}}$$

In equation (14), the right hand side terms are known where as those in left hand are unknown and are functions of water depth. For a few commonly encountered sections the parameters A and R are given in the table below.

	Rectangle	Trapezoid	Circle
Flow Area, A	$b.h$	$(b+my).y$	$\frac{1}{8}(\phi - sin\phi)$
Wetted Perimeter, P	$b+2h$	$b+2h\sqrt{1+m^2}$	$\frac{1}{2}\phi D$
Hydraulic Radius, R	$\frac{bh}{b+2h}$	$\frac{(b+my).y}{b+2h\sqrt{1+m^2}}$	$\frac{1}{4}\left(1-\frac{\sin\phi}{\phi}\right)D$
Free surface width, B	b	$b+2mh$	$\left(sin\frac{\phi}{2}\right)D$

In the table, m stands for the side slope of a trapezoidal channel and < stands for the angle subtended at the centre by the water surface chord line.

As seen from the above table except for the very simple rectangular section it is not possible directly to evaluate h, corresponding to Q as the left hand side of equation earlier is nonlinear in terms of h. One way of solving is by Newton's method, where equation above is written as,

$$f(h) = AR^{2/3} - \frac{nQ}{S^{1/2}} = 0$$

For using Newton's method the derivative of the function is required,

$$f'(h) = \frac{d}{dh}\left(A\frac{A^{2/3}}{P^{2/3}} - \frac{nQ}{S^{1/2}}\right) = 0$$

$$f'(h) = \frac{5}{3}P^{-2/3}A^{2/3}\frac{dA}{dh} - \frac{2}{3}P^{-5/3}A^{5/3}\frac{dP}{dh}$$

$$f'(h) = \frac{5}{3}BR^{2/3} - \frac{2}{3}R^{2/3}\frac{dP}{dh}$$

Where we have used $\frac{dA}{dh} = B$. Similarly the expression $\frac{dP}{dh}$ may be evaluated for any section.

Starting with a realistic value h_i the iteration may be carried out as given below:

$$h^{i+1} = h^i - \frac{f(h^i)}{f'(h^i)}$$

Where h^{i+1} is the value of h at next iteration, which is an improvement of initial guess h^i. The iteration may be continued till a desired accuracy is achieved.

Uniform Flow in Channels of Compound Cross Section

A compound section may be defined as a section in which various portions of the cross-section have different flow properties, like surface roughness or channel depth. (Figure o).

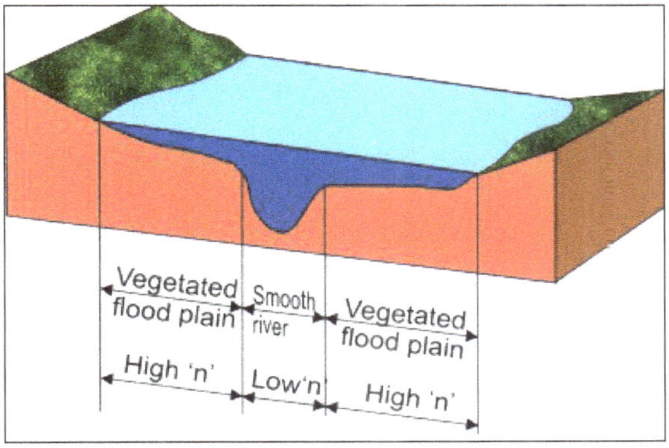

FIGURE o. A river in flood example of a compound flow section.

In order to use the uniform flow formula in compound channels one way may be to divide the flow section into sub areas (Figure p) and treat the flow in each area separately.

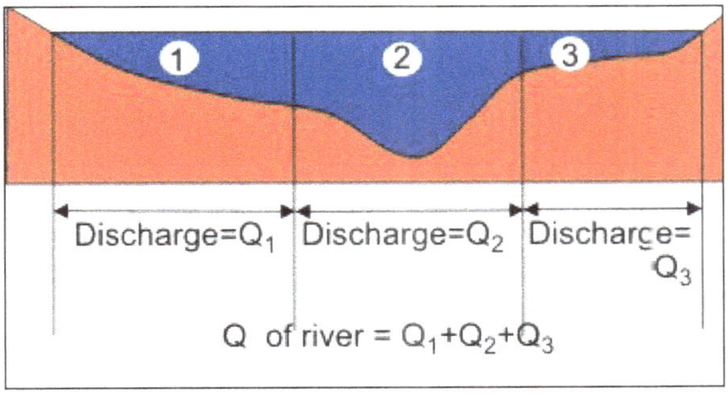

FIGURE p. Simple flow analysis method for compound channels.

However, it has been found that this method may lead to errors by as much as ±20% or even more (Chadwick et al 2004). The error is largely due to the neglecting of mass and momentum interchange between adjacent sub-areas. The current solution would however be more complex by using a two or even three-dimensional model.

In another method, the energy coefficient (α) and friction slope S_f are evaluated in terms of conveyance K of the sub areas. With these expressions, the flow in compound section may be computed without knowing the individual flows in each sub area. For a compound channel divided into N sections. The energy coefficient, α, is found out as:

$$\alpha = \frac{\sum_{i=1}^{N} V_i^3 A_i}{V_m^3 \sum_{i=1}^{N} A_i}$$

Where V_m is the mean flow velocity in the entire section and is given as follows,

$$\frac{\sum V_i A_i}{\sum}$$

Where $V_i = Q_i / A_i$ and A_i is the area of its ith sub-area. Equation (18) now can be written as,

$$\alpha = \frac{\left(\sum Q_i^3 / A_i^2\right)\left(\sum A_i\right)^2}{\left(\sum Q_i\right)^3}$$

Now, the flow in sub-areas i may be written as,

$$Q_i = K_i S_{fi}^{1/2}$$

$$S_{fi}^{1/2} = \frac{Q_i}{K_i}$$

Here, an assumption has been made that S_f has the same value for all sub- areas, which is not quite correct since the velocities of each of these areas being different, would not give equal velocity heads. Where as, the water surface is almost level over the entire cross section.

$$\frac{Q_1}{K_1} = \frac{Q_2}{K_2} = - - - - - - \frac{Q_n}{K_n} = \text{Constant} = S_f^{1/2}$$

It follows from equation (23) that,

$$Q_1 = K_1 \frac{Q_n}{K_n}$$

$$Q_2 = K_2 \frac{Q_n}{K_n}$$

–

–

–

$$Q_n = K_n \frac{Q_n}{K_n}$$

Adding all the above equation yields,

$$Q = \sum_{i=1}^{n} Q_i = \frac{Q_n}{K_n} \sum_{i=1}^{n} K_i$$

By substituting this expression for $Q_i = K_i \left(\frac{Q_n}{K_n} \right)$ into equation above and simplifying the equation, one obtains,

$$\alpha = \frac{\left(\sum_{i=1}^{n} A_i \right)^2}{\left(\sum_{i=1}^{n} K_i \right)^3}, \sum_{i=1}^{n} \left(\frac{K_i^3}{A_i^2} \right)$$

Elimination of $\frac{Q_n}{K_n}$ from equations above and squaring both sides give,

$$S_f = \left(\frac{\sum Q_i}{\sum K_i} \right)^2$$

$$S_f = \frac{Q^2}{\sum K_i^2}$$

Thus, expressions for α and S_f have been evaluated for any given stage without explicitly determining the flow in each sub areas, Q_i. In addition, equation (30) may be used in the procedure for determining varied flow profiles.

Non Uniform in Channels

There are quite a few examples of non-uniform flow in rivers or open channels that may be encountered by a water resources engineer. Some of these have been illustrated in Figure q.

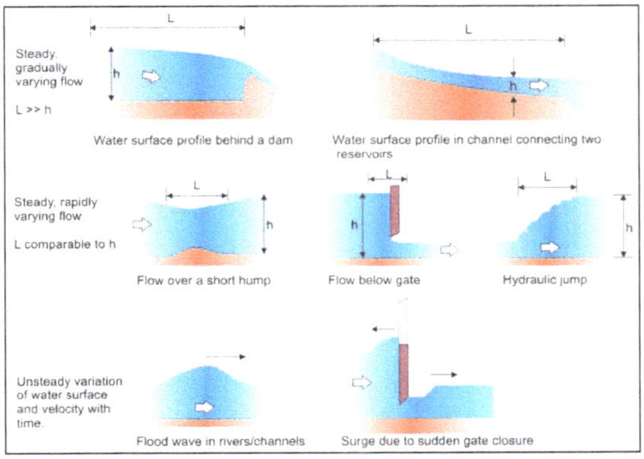

FIGURE q. Some example of flow situation encountered in practice.

In this lesson we shall discuss the procedure to evaluate water surface profiles for steady, gradually varying flow situations.

Non-uniform Gradually Varied Flow Calculation

A representative non-uniform gradually varied flow is shown in Figure r.

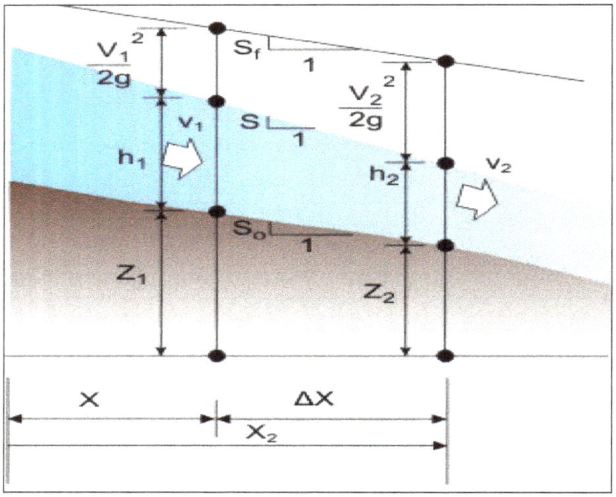

FIGURE r. Non-uniform gradually varied flow.

Over the incremental distance Δx, the depth and velocity are known to change slowly. The slope of the energy grade line is designated as α in contrast to uniform flow, the slopes of the energy grade line, water surface, and channel bottom are no longer parallel. Since the changes in the water depth h and velocity V are gradual, the energy lost over the incremental Δx can be represented by manning equation. This means that equation, which is valid for uniform flow can also be used to evaluate S for a gradual varied flow situation, and that the roughness coefficients discussed.

Additional assumption includes a regular cross section, small channel slope, hydrostatic pressure distribution and one-dimensional flow.

Applying the equivalence of energy between locations 1 and 2, and assuming the loss term as h_L given by $S_f \cdot \Delta x$ one obtains,

$$Z_1 + h_1 + \alpha \frac{V_1^2}{2g} = Z_2 + h_2 + \alpha \frac{V_2^2}{2g} + S_f \Delta x$$

In the above equation, Δx is the distance between two consecutive sections x_1 and x_2 such that $\Delta x = x_2 - x_1$. The energy coefficient α has been used along with the $\frac{V^2}{2g}$ term, as it may be much different from 1.0 for natural sections. The term S_f in equation above may be evaluated by the expression for uniform flow, equation above, where S_o may be replaced by S_f. Since equation above relates the energy between the sections, S_f may be taken either of the following:

$$\text{Arithmetic mean: } \overline{S_f} = \frac{1}{2}\left(S_{f1} + S_{f2}\right)$$

$$\text{Geometric mean: } \overline{S_f} = \sqrt{S_{f1}\, S_{f2}}$$

$$\text{Harmonic mean: } \overline{S_f} = \frac{2S_{f1}\, S_{f2}}{S_{f1}\ \ S_{f2}}$$

Where S_{f1} and S_{f2} are the friction slopes evaluated at section 1 and 2 by using the Mannings formula equation earlier.

Equation earlier may be used by starting from one end of the channel where the flow depth and velocity are known and working backward or forward in steps. Here, two, methods are used of which we shall discuss one, called the standard step method. Every popular computer program called HEC-2 developed by hydrologic engineering center of the US Army Corps of Engineers is based on this method.

In the standard step method, for any given discharge the depth of flow would be known at the control section. It is then required to calculate the depth of flow at the section immediately next to the control section. Two examples are illustrated in Figure s.

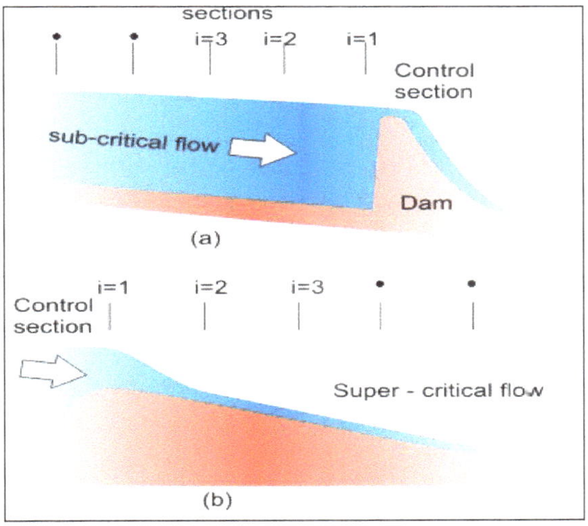

FIGURE s. Computation grid by the standard step method.

(a) profile behind a dam: calculation proceeds upstream from control section.

(b) profile in a steep channel : calculation proceeds downstream from control section.

The distance between the two successive sections (i and i+1) is taken as constant, say Δx. It may be observed from the Figure 19a since the water is flowing above the dam the water depth above the dam crest can be found out for the given discharge. Hence the water level at the control section just upstream of the dam is known. Similarly, in Figure 19b, since the water is flowing down from the reservoir into the steep channel critical depth corresponding to the given discharge would exist at the control section. Here two, the water level at the control section is then known.

Starting at the control section (i =1), the total energy of water is found out to be,

$$H_1 = Z_1 + h_1 + \alpha \frac{V_1^2}{2g}$$

Next, consider the first reach, that is, between sections i =1 and i =2. A depth of flow is assumed at section 2 and the energy there, that is,

$$H_2 = Z_2 + h_2 + \alpha \frac{V_2^2}{2g}$$

is evaluated. Now, one of the equations for finding $\overline{S_f}$ (the average friction slope) in the reach is found out by, say equation earlier.

As may be observed from Figure r the numerical value of H_2 found from equation above should be equal to that of h_1 found from equation $+S_f$. If the depth at the section 2 has been correctly assumed if the two don't match, a new depth h_2 is assumed and the calculations are repeated till the two values match.

Once a correct depth is found at section 2, a similar procedure is used to find the depth at section 3, and so on.

These are the two other methods to find out water surface profiles of gradually varied flow situations, namely; method of direct integration and method of graphical integration.

Gradually Varied Flow Profiles

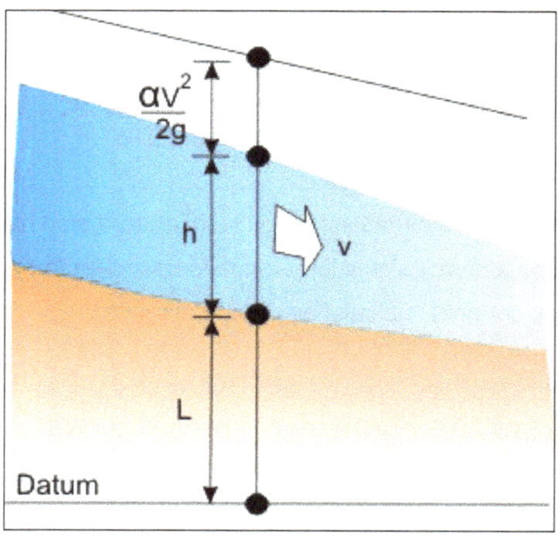

FIGURE t. Definition sketch.

In many flow problems it is enough to make a qualitative sketch of water surface profile for a given flow that is taking place between two locations. It is not necessary therefore to find out the exact level of water at different points but the general shape of the free surface has to be drawn as accurately as possible. An analysis of water surface profile may be done by studying the governing equation, which can be derived from the sketch in Figure t.

The total energy H at a channel section is given as,

$$\mathrm{H} = Z + h + \alpha \frac{V^2}{2g}$$

Where

- H: Elevation of energy line above the datum.
- Z: Elevation of channel bottom above datum.
- h: Flow depth.
- V: Mean flow velocity.
- α : Velocity head coefficient.

Considering x as the space coordinate, taken positive in the direction of flow one obtains by differentiating both sides of the equation above with respect to x and expressing V in terms of discharge Q.

$$\frac{dH}{dx} = \frac{dZ}{dx} + \frac{dh}{dx} + \alpha \frac{Q^2}{2g} \frac{d}{dx}\left(\frac{1}{A^2}\right)$$

Again, we know by definition:

$$\frac{dH}{dx} = -S_f$$

$$\text{And } \frac{dZ}{dx} = -S_o$$

In which

- S_f: Slope of the energy grade line.
- S_o: Slope of the channel bottom.

The negative sign of S_f and S_o indicates that both H and Z decrease as x increases. In equation above an expression for the derivative of A^{-2} may be found out as follows,

$$\frac{d}{dx}\left(\frac{1}{A^2}\right)=\frac{d}{dA}\left(\frac{1}{A^2}\right)\frac{dA}{dx}$$
$$=\frac{d}{dA}\left(\frac{1}{A^2}\right)\frac{dA}{dh}\frac{dh}{dx}$$
$$=-\frac{2B}{A^3}\frac{dh}{dx}$$

Since $\frac{dA}{dh}=B$

By substituting equations earlier, and rearranging the resulting equation one obtains,

$$\frac{dh}{dx}=\frac{S_o-S_f}{1-\left(\alpha BQ^2\right)/gA^3}$$

If the channel is not prismatic, then the cross sectional area A changes with distance, and may be expressed as:

$$\frac{dA}{dx}=\frac{\partial A}{\partial x}+\frac{\partial A}{\partial y}\frac{dh}{dx}$$

The above change would modify equations earlier accordingly.

We may express equation, which describes the variation of h with x, in terms of the Froude Number (Fr) if we note the following:

$$\frac{\alpha BQ^2}{gA^3}=\frac{(Q/A)^2}{(gA)/(\alpha B)}=Fr^2$$

Hence, equation may be written as,

$$\frac{dh}{dx}=\frac{S_o-S_f}{1-Fr^2}$$

Equation above can give a general idea about the nature of the curve if one knows the relative inclinations of the channel bed slope and friction slope (S_f) and the Froude Number (Fr). This may be done by observing the water flow depth (h) with respect to normal depth (h_n) and critical depth (h_c) for a given discharge, the following figures show the relative changes of h_n and h_c as channel bed slope is increased gradually from horizontal. It may be observed that the for a given discharge h_c does not change but h_n goes on decreasing starting from an infinite value for a flat slope.

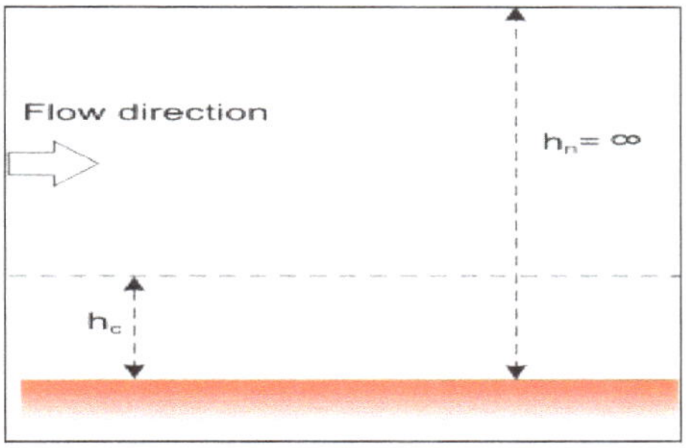

FIGURE u. Normal depth (h_n) and critical depth (h_c) on horizontal sloped bed.

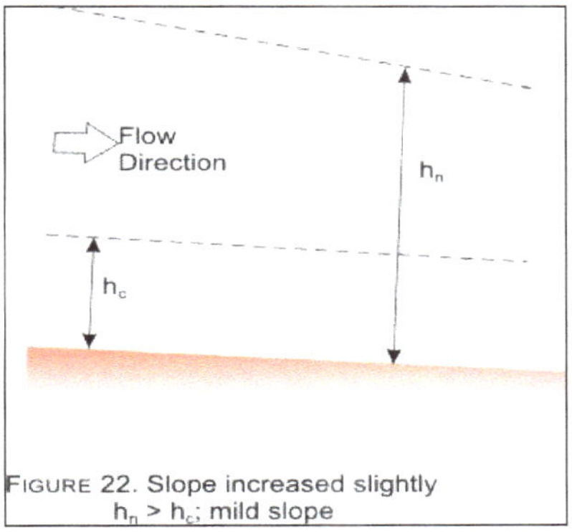

FIGURE v. Slope increased slightly $h_n > h_c$; mild slope.

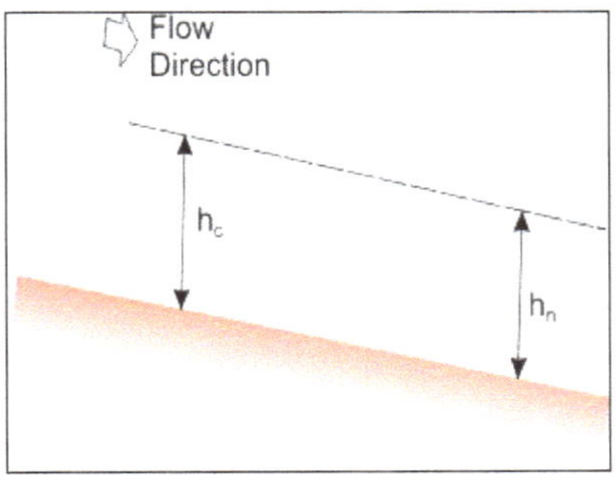

FIGURE w. Slope increased further $h_n > h_c$; critical slope.

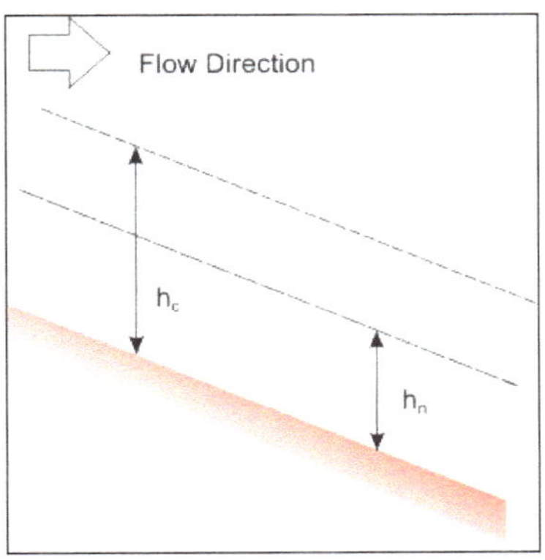

FIGURE x. Slope increased substantially $h_n > h_c$; Steep slope.

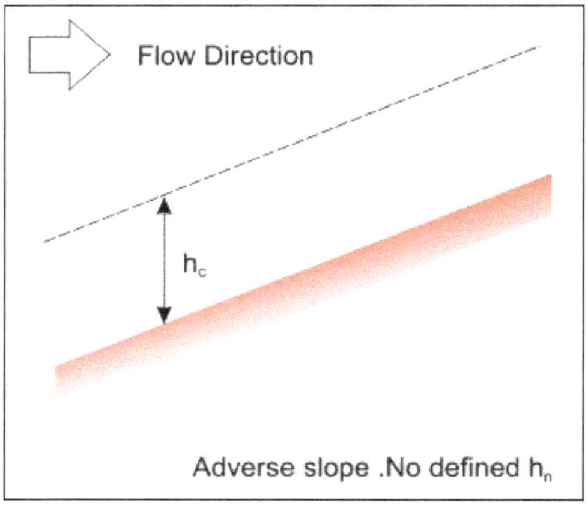

FIGURE y. Adverse slope. No defined h_n

In water resources projects, one generally encounters slopes of channels that are either of the following:

- Mild, where $h_n > h_c$ (Figure v)
- Steep, where $h_n < h_c$ (Figure x)
- Critical, where $h_n = h_c$(Figure w)
- Flat, where $h_n = \infty$ (Figure u)
- Adverse, where the slope is reversed (Figure y)

For each of these slopes, the actual water surface would vary depending upon a control that exist either at the upstream or downstream end of the channel. Some examples of controls are given below,

- Weir or spillway (Figures z and a1)
- Gate (Figure a2)
- Free overfall (Figure a3)

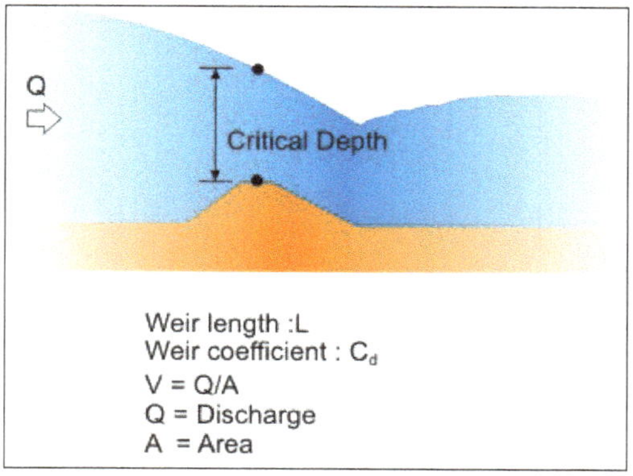

FIGURE z. Flow over a weir.

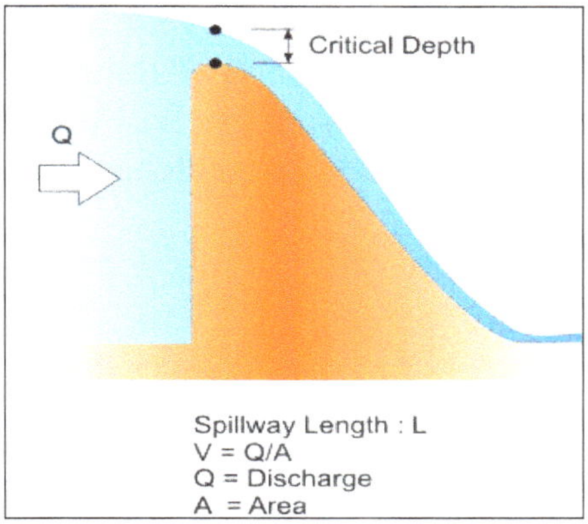

FIGURE a1. Flow over a dam spilway.

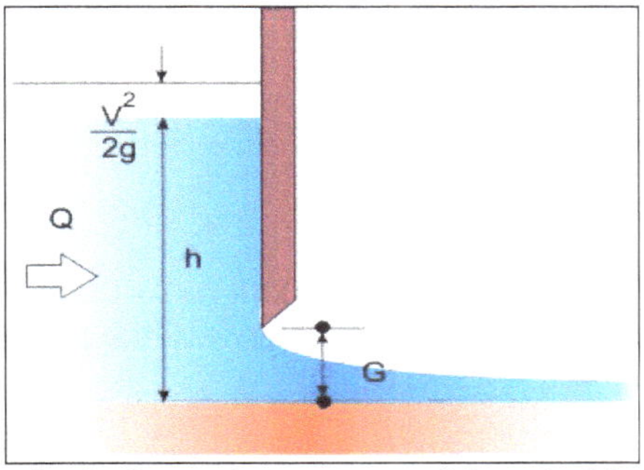

FIGURE a2. Flow below a gate. The depth of water just upstream of the gate (h) may be determined from the formula $Q=C.L.G\sqrt{H}$ where $H=h+v^2/2g$. The formula is valid for small gate openings.

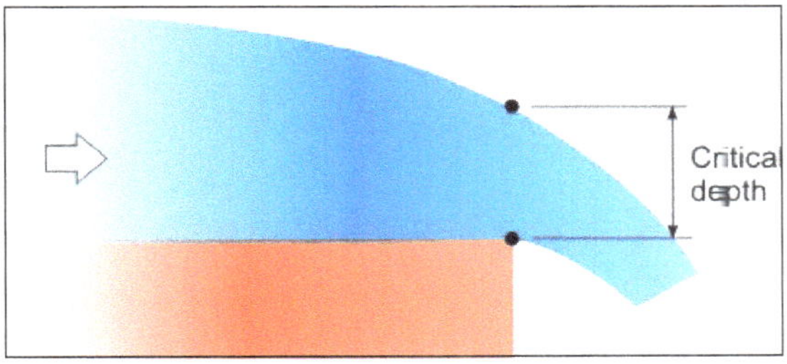

FIGURE a3. Free over fall.

Apart from the above a normal depth may be assumed to exist within a very long channel, for which the conditions at the far end may be neglected (Figure a4).

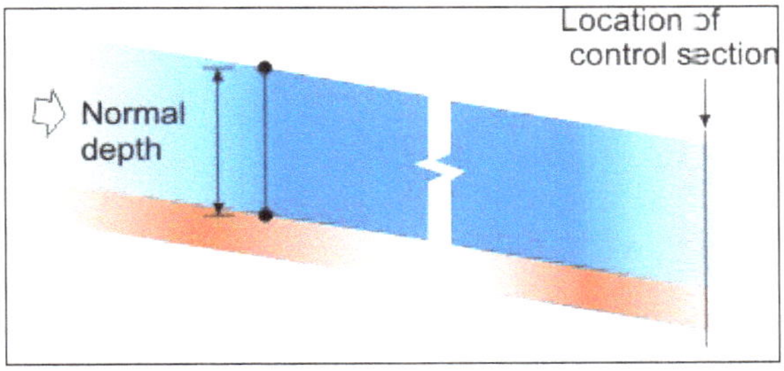

FIGURE a4. Flow in a long channel.

The situation shown in Figure a4 is used often while analyzing flow in, say, at the tail end of long irrigation channels or in a long river. Examples illustrating the use of equation and a known control section in determining flow profiles where for a mildly slope channel. Similar profiles may be qualitatively sketched for other channels too.

Downstream Control Raising the Water Level Above Normal Depth

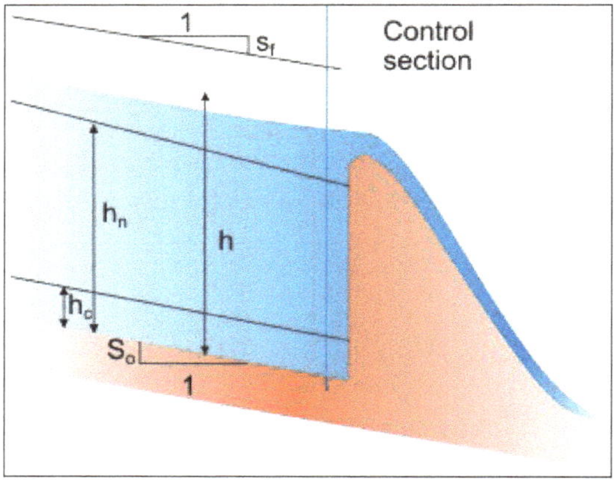

FIGURE a5. M1 type of water surface profile behind dam spillway.

This situation is common for spillways of large dams. The flow profile in a mildly sloped channels where $h > h_n > h_c$ as shown in Figure 31 is known as the M_1 curve. Now, for uniform flow, $S_f = S = S_o$ when $h = h_n$. Hence it is clear from Mannings formula (equation 11), that for a given discharge, Q,

$$S_o > S_f \text{ if } h > h_n$$

Thus, in equation,

$$\frac{dh}{dx} = \frac{S_o - S_f}{1 - Fr^2}$$

The numerator is positive Fr<1 since h>h_c. Therefore, the denominator of equation above is positive as well. Hence, it follows from this eqn that,

$$\frac{dh}{dx} = \frac{S_o - S_f}{1 - Fr^2} = \frac{+}{+} = +$$

This means that h increases with distance x.

Comparing with Figure a3 it may be inferred that quite some distance upstream of the spillway the flow depth nearly equals normal depth. And, since dh/dx for this profile is positive which means that the water depth goes on increasing towards the spillway, the flow depth becomes nearly horizontal. However very close to spillway the flow profile again changes which is due to the fact that the flow here is not really one-dimensional (Figure a6).

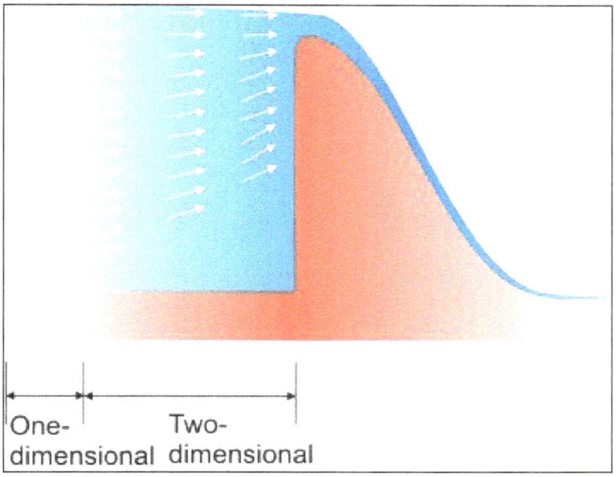

FIGURE a6. Flow situation very close to a spillway

Downstream Controlled Raising Water Level Above Critical Depth but Below Normal Depth

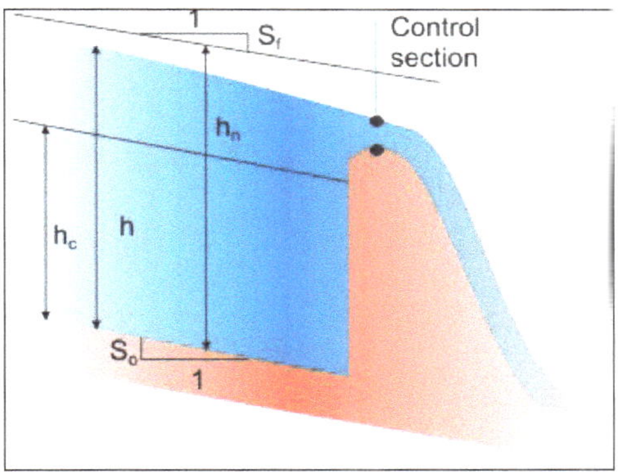

FIGURE a7. M2 type of water surface profile behind a short-heigt spillway.

The flow profile in a mildly sloping channel where $h_n>h>h_c$, has been shown in Figure a7 is known as the M_2 curve. In this case $S_f>S_o$ since $h<h_n$ (from Mannings formula). Thus the numerator in equation above is negative. However, the denominator is positive, since Fr<1 because $h>h_c$ hence it follows from equation above that,

$$\frac{dh}{dx}=\frac{S_o-S_f}{1-Fr^2}=\frac{-}{-}=-$$

Thus h decreases as x increases for upstream of this spillway control section the flow depth would be asymptotic to normal depth h_n.

Upstream Control Causing Water Depth to be Less than Both Normal and Critical Depths

This situation is shown in Figure a8 for flow taking place below a sluice gate. The reader is advised to check the trend of water surface profile using equation above in this case.

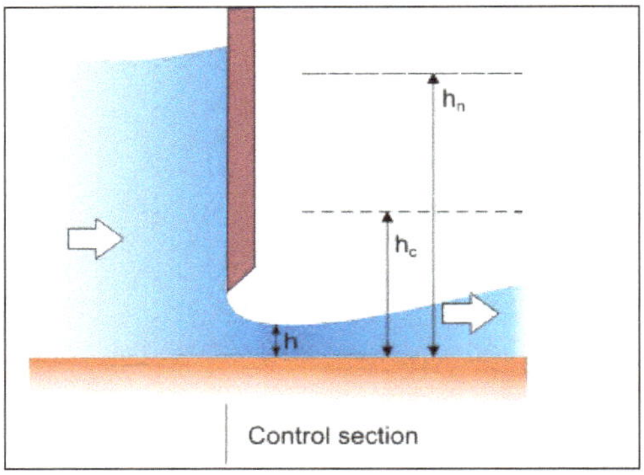

FIGURE a8. M3 type of water surface profile downstream of a gate.

Wastewater

Greywater (a type of wastewater) in a settling tank.

Wastewater, also written as waste water, is any water that has been adversely affected in quality by anthropogenic influence. Wastewater can originate from a combination of domestic, industrial, commercial or agricultural activities, surface runoff or stormwater, and from sewer inflow or infiltration.

Municipal wastewater (also called sewage) is usually conveyed in a combined sewer or sanitary sewer, and treated at a wastewater treatment plant. Treated wastewater is discharged into receiving water via an effluent pipe. Wastewaters generated in areas without access to centralized sewer systems rely on on-site wastewater systems. These typically comprise a septic tank, drain field, and optionally an on-site treatment unit. The management of wastewater belongs to the overarching term sanitation, just like the management of human excreta, solid waste and stormwater (drainage).

Sewage is a type of wastewater that comprises domestic wastewater and is therefore contaminated with feces or urine from people's toilets, but the term sewage is also used to mean any type of wastewater. Sewerage is the physical infrastructure, including pipes, pumps, screens, channels etc. used to convey sewage from its origin to the point of eventual treatment or disposal.

Sources

Wastewater can come from:

- Human excreta (feces and urine) often mixed with used toilet paper or wipes; this is known as blackwater if it is collected with flush toilets.
- Washing water (personal, clothes, floors, dishes, cars, etc.), also known as greywater or sullage.
- Surplus manufactured liquids from domestic sources (drinks, cooking oil, pesticides, lubricating oil, paint, cleaning liquids, etc.)
- Urban rainfall runoff from roads, carparks, roofs, sidewalks/pavements (contains oils, animal feces, litter, gasoline/petrol, diesel or rubber residues from tires, soapscum, metals from vehicle exhausts, etc.)
- Highway drainage (oil, de-icing agents, rubber residues, particularly from tires)·
- Storm drains (may include trash).
- Manmade liquids (illegal disposal of pesticides, used oils, etc.)
- Industrial waste.
- Industrial site drainage (silt, sand, alkali, oil, chemical residues);
    - Industrial cooling waters (biocides, heat, slimes, silt).
    - Industrial process waters.
    - Organic or biodegradable waste, including waste from abattoirs, creameries, and ice cream manufacture.
    - Organic or non bio-degradable/difficult-to-treat waste (pharmaceutical or pesticide manufacturing).

- o Extreme pH waste (from acid/alkali manufacturing, metal plating).
- o Toxic waste (metal plating, cyanide production, pesticide manufacturing, etc.)
- o Solids and emulsions (paper manufacturing, foodstuffs, lubricating and hydraulic oil manufacturing, etc.)
- o Agricultural drainage, direct and diffuse.
- o Hydraulic fracturing.
- o Produced water from oil & natural gas production.

Wastewater can be diluted or mixed with other types of water in the form of:

- Seawater ingress (high volumes of salt and microbes).
- Direct ingress of river water.
- Rainfall collected on roofs, yards, hard-standings, etc. (generally clean with traces of oils and fuel).
- Groundwater infiltrated into sewage.

After it has undergone some treatment, the "treated wastewater" remains, e.g.:

- Septic tank discharge.
- Sewage treatment plant discharge.

Constituents

The composition of wastewater varies widely. This is a partial list of what it may contain:

- Water (more than 95 percent), which is often added during flushing to carry waste down a drain;
- Pathogens such as bacteria, viruses, prions and parasitic worms;
- Non-pathogenic bacteria;
- Organic particles such as feces, hairs, food, vomit, paper fibers, plant material, humus, etc.;
- Soluble organic material such as urea, fruit sugars, soluble proteins, drugs, pharmaceuticals, etc.;
- Inorganic particles such as sand, grit, metal particles, ceramics, etc.;
- Soluble inorganic material such as ammonia, road-salt, sea-salt, cyanide, hydrogen sulfide, thiocyanates, thiosulfates, etc.;

- Animals such as protozoa, insects, arthropods, small fish, etc.;
- Macro-solids such as sanitary napkins, nappies/diapers, condoms, needles, children's toys, dead animals or plants, etc.;
- Gases such as hydrogen sulfide, carbon dioxide, methane, etc.;
- Emulsions such as paints, adhesives, mayonnaise, hair colorants, emulsified oils, etc.;
- Toxins such as pesticides, poisons, herbicides, etc.
- Pharmaceuticals and hormones and other hazardous substances.

Quality Indicators

Any oxidizable material present in an aerobic natural waterway or in an industrial wastewater will be oxidized both by biochemical (bacterial) or chemical processes. The result is that the oxygen content of the water will be decreased. Basically, the reaction for biochemical oxidation may be written as:

Oxidizable material + bacteria + nutrient + $O_2 \rightarrow CO_2 + H_2O$ + oxidized inorganics such as NO_3^- or SO_4^{2-}

Oxygen consumption by reducing chemicals such as sulfides and nitrites is typified as follows:

$$S^{2-} + 2\,O_2 \rightarrow SO_4^{2-}$$

$$NO_2^- + \tfrac{1}{2}\,O_2 \rightarrow NO_3^-$$

Since all natural waterways contain bacteria and nutrients, almost any waste compounds introduced into such waterways will initiate biochemical reactions (such as shown above). Those biochemical reactions create what is measured in the laboratory as the biochemical oxygen demand (BOD). Such chemicals are also liable to be broken down using strong oxidizing agents and these chemical reactions create what is measured in the laboratory as the chemical oxygen demand (COD). Both the BOD and COD tests are a measure of the relative oxygen-depletion effect of a waste contaminant. Both have been widely adopted as a measure of pollution effect. The BOD test measures the oxygen demand of biodegradable pollutants whereas the COD test measures the oxygen demand of oxidizable pollutants.

The so-called 5-day BOD measures the amount of oxygen consumed by biochemical oxidation of waste contaminants in a 5-day period. The total amount of oxygen consumed when the biochemical reaction is allowed to proceed to completion is called the Ultimate BOD. Because the Ultimate BOD is so time consuming, the 5-day BOD has been almost universally adopted as a measure of relative pollution effect.

There are also many different COD tests of which the 4-hour COD is probably the most common.

There is no generalized correlation between the 5-day BOD and the ultimate BOD. Similarly there is no generalized correlation between BOD and COD. It is possible to develop such correlations for specific waste contaminants in a specific wastewater stream but such correlations cannot be generalized for use with any other waste contaminants or wastewater streams. This is because the composition of any wastewater stream is different. As an example an effluent consisting of a solution of simple sugars that might discharge from a confectionery factory is likely to have organic components that degrade very quickly. In such a case, the 5 day BOD and the ultimate BOD would be very similar since there would be very little organic material left after 5 days. However a final effluent of a sewage treatment works serving a large industrialised area might have a discharge where the ultimate BOD was much greater than the 5 day BOD because much of the easily degraded material would have been removed in the sewage treatment process and many industrial processes discharge difficult to degrade organic molecules.

The laboratory test procedures for the determining the above oxygen demands are detailed in many standard texts. American versions include the "Standard Methods for the Examination of Water and Wastewater."

Treatment

There are numerous processes that can be used to clean up wastewaters depending on the type and extent of contamination. Wastewater can be treated in wastewater treatment plants which include physical, chemical and biological treatment processes. Municipal wastewater is treated in sewage treatment plants (which may also be referred to as wastewater treatment plants). Agricultural wastewater may be treated in agricultural wastewater treatment processes, whereas industrial wastewater is treated in industrial wastewater treatment processes.

For municipal wastewater the use of septic tanks and other On-Site Sewage Facilities (OSSF) is widespread in some rural areas, for example serving up to 20 percent of the homes in the U.S.

One type of aerobic treatment system is the activated sludge process, based on the maintenance and recirculation of a complex biomass composed of micro-organisms able to absorb and adsorb the organic matter carried in the wastewater. Anaerobic wastewater treatment processes (UASB, EGSB) are also widely applied in the treatment of industrial wastewaters and biological sludge. Some wastewater may be highly treated and reused as reclaimed water. Constructed wetlands are also being used.

Disposal

In some urban areas, municipal wastewater is carried separately in sanitary sewers and runoff from streets is carried in storm drains. Access to either of these is typically through a manhole. During high precipitation periods a combined sewer overflow can occur, forcing untreated sewage to flow back into the environment. This can pose a serious threat to public health and the surrounding environment.

Industrial wastewater effluent with neutralized pH from tailing runoff in Peru.

Sewage may drain directly into major watersheds with minimal or no treatment but this usually has serious impacts on the quality of an environment and on the health of people. Pathogens can cause a variety of illnesses. Some chemicals pose risks even at very low concentrations and can remain a threat for long periods of time because of bioaccumulation in animal or human tissue.

Wastewater may be pumped underground through an injection well.

Reuse

Treated wastewater can be reused in industry (for example in cooling towers), in artificial recharge of aquifers, in agriculture and in the rehabilitation of natural ecosystems (for example in Florida's Everglades). In rarer cases it is also used to augment drinking water supplies.

There are several technologies used to treat wastewater for reuse. A combination of these technologies can meet strict treatment standards and make sure that the processed water is hygienically safe, meaning free from bacteria and viruses. The following are some of the typical technologies: Ozonation, ultrafiltration, aerobic treatment (membrane bioreactor), forward osmosis, reverse osmosis, advanced oxidation.

Some water demanding activities do not require high grade water. In this case, wastewater can be reused with little or no treatment. One example of this scenario is in the domestic environment where toilets can be flushed using greywater from baths and showers with little or no treatment.

Gardening and Agriculture

There are benefits of using recycled water for irrigation, including the lower cost compared to some other sources and consistency of supply regardless of season, climatic conditions and associated water restrictions. Irrigation with recycled wastewater can also serve to fertilize plants if it contains nutrients, such as nitrogen, phosphorus and potassium.

Around 90% of wastewater produced globally remains untreated, causing widespread water pollution, especially in low-income countries. Increasingly, agriculture is using untreated wastewater for irrigation. Cities provide lucrative markets for fresh produce, so are attractive to farmers. However, because agriculture has to compete for increasingly scarce water resources with industry and municipal users, there is often no alternative for farmers but to use water polluted with urban waste directly to water their crops.

Health Risks

There can be significant health hazards related to using untreated wastewater in agriculture. Wastewater from cities can contain a mixture of chemical and biological pollutants. In low-income countries, there are often high levels of pathogens from excreta, while in emerging nations, where industrial development is outpacing environmental regulation, there are increasing risks from inorganic and organic chemicals. The World Health Organization, in collaboration with the Food and Agriculture Organization of the United Nations (FAO) and the United Nations Environmental Program (UNEP), has developed guidelines for safe use of wastewater in 2006. These guidelines advocate a 'multiple-barrier' approach to wastewater use, for example by encouraging farmers to adopt various risk-reducing behaviours. These include ceasing irrigation a few days before harvesting to allow pathogens to die off in the sunlight, applying water carefully so it does not contaminate leaves likely to be eaten raw, cleaning vegetables with disinfectant or allowing fecal sludge used in farming to dry before being used as a human manure.

Legislation

European Union

Council Directive 91/271/EEC on Urban Wastewater Treatment was adopted on 21 May 1991, amended by the Commission Directive 98/15/EC. Commission Decision 93/481/EEC defines the information that Member States should provide the Commission on the state of implementation of the Directive.

United States

The Clean Water Act is the primary federal law in the United States governing water pollution.

Philippines

In the Philippines, Republic Act 9275, otherwise known as the Philippine Clean Water Act of 2004, is the governing law on wastewater management. It states that it is the country's policy to protect, preserve and revive the quality of our fresh, brackish and marine waters, for which wastewater management plays a particular role.

References

- "Texas fracking site that spilled 42,000 gallons of fluid into residential area hopes to reopen". RT International. Retrieved 14, January 2020
- World Health Organization (WHO) (2006). "Protecting Groundwater for Health - Understanding the drinking-water catchment" (PDF). Retrieved 07, March 2020
- G. Allen Burton, Jr., Robert Pitt (2001). Stormwater Effects Handbook: A Toolbox for Watershed Managers, Scientists, and Engineers. New York: CRC/Lewis Publishers. ISBN 0-87371-924-7
- Goel, P.K. (2006). Water Pollution - Causes, Effects and Control. New Delhi: New Age International. p. 179. ISBN 978-81-224-1839-2
- ATSDR (US Agency for Toxic Substance & Disease Registry) (2008). "Follow-up Health Consultation: Anniston Army Depot." (PDF). Retrieved 18, June 2020
- Suthar, S; Bishnoi, P; Singh, S; et al. (2009). "Nitrate contamination in groundwater of some rural areas of Rajasthan, India". Journal of Hazardous Materials. 171 (1–3): 189–199. doi:10.1016/j.jhazmat.2009.05.111
- Smith, M; Cross, K; Paden, M; Laben, P, eds. (2016). Spring - managing groundwater sustainably (PDF). IUCN. ISBN 978-2-8317-1789-0
- Laws, Edward A. (2000). Aquatic Pollution: An Introductory Text. New York: John Wiley and Sons. p. 430. ISBN 978-0-471-34875-7
- "Report Offers First Worldwide Estimate of Investments in Combating Water Pollution". American Association for the Advancement of Science. Retrieved 01, August 2020
- Wachman, Richard (2007-12-09). "Water becomes the new oil as world runs dry". The Guardian. London. Retrieved 15, July 2020
- Pfunt,H; Houben,G; Himmelsbach,T (2016). "Numerical modeling of fracking fluid migration through fault zones and fractures in the North German Basin". Hydrogeology Journal. 24 (6): 1343–1358. doi:10.1007/s10040-016-1418-7

5

Strategies for Irrigation

Irrigation is an artificial process of supplying water to land or crops to assist in the production of crops. A few of the strategies studied in relation to irrigation are deficit irrigation, flexible barge, hippo water roller and peak water. This chapter closely examines the key concepts of irrigation to provide an extensive understanding of the subject.

Deficit Irrigation

Deficit irrigation (DI) is a watering strategy that can be applied by different types of irrigation application methods. The correct application of DI requires thorough understanding of the yield response to water (crop sensitivity to drought stress) and of the economic impact of reductions in harvest. In regions where water resources are restrictive it can be more profitable for a farmer to maximize crop water productivity instead of maximizing the harvest per unit land. The saved water can be used for other purposes or to irrigate extra units of land. DI is sometimes referred to as incomplete supplemental irrigation or regulated DI.

Definition

Deficit irrigation (DI) has been reviewed and defined as follows:

"Deficit irrigation is an optimization strategy in which irrigation is applied during drought-sensitive growth stages of a crop. Outside these periods, irrigation is limited or even unnecessary if rainfall provides a minimum supply of water. Water restriction is limited to drought-tolerant phenological stages, often the vegetative stages and the late ripening period. Total irrigation application is therefore not proportional to irrigation requirements throughout the crop cycle. While this inevitably results in plant drought stress and consequently in production loss, DI maximizes irrigation water productivity, which is the main limiting factor (English, 1990). In other words, DI aims at stabilizing yields and at obtaining maximum crop water productivity rather than maximum yields (Zhang and Oweis, 1999)."

Crop Water Productivity

Crop water productivity (WP) or water use efficiency (WUE) expressed in kg/m^3 is an efficiency term, expressing the amount of marketable product (e.g. kilograms of grain) in relation to the amount of input needed to produce that output (cubic meters of water). The water used for crop production is referred to as crop evapotranspiration. This is a combination of water lost by evaporation from the soil surface and transpiration by the plant, occurring simultaneously. Except by modeling, distinguishing between the two processes is difficult. Representative values of WUE for cereals at field level, expressed with evapotranspiration in the denominator, can vary between 0.10 and 4 kg/m3.

Experiences with Deficit Irrigation

For certain crops, experiments confirm that DI can increase water use efficiency without severe yield reductions. For example for winter wheat in Turkey, planned DI increased yields by 65% as compared to winter wheat under rainfed cultivation, and had double the water use efficiency as compared to rainfed and fully irrigated winter wheat. Similar positive results have been described for cotton. Experiments in Turkey and India indicated that the irrigation water use for cotton could be reduced to up to 60 percent of the total crop water requirement with limited yield losses. In this way, high water productivity and a better nutrient-water balance was obtained.

Certain Underutilized and horticultural crops also respond favorably to DI, such as tested at experimental and farmer level for the crop quinoa. Yields could be stabilized at around 1.6 tons per hectare by supplementing irrigation water if rainwater was lacking during the plant establishment and reproductive stages. Applying irrigation water throughout the whole season (full irrigation) reduced the water productivity. Also in viticulture and fruit tree cultivation, DI is practiced.

Scientists affiliated with the Agricultural Research Service (ARS) of the USDA found that conserving water by forcing drought (or deficit irrigation) on peanut plants early in the growing season has shown to cause early maturation of the plant yet still maintain sufficient yield of the crop. Inducing drought through deficit irrigation earlier in the season caused the peanut plants to physiologically "learn" how to adapt to a stressful drought environment, making the plants better able to cope with drought that commonly occurs later in the growing season. Deficit irrigation is beneficial for the farmers because it reduces the cost of water and prevents a loss of crop yield (for certain crops) later on in the growing season due to drought. In addition to these findings, ARS scientists suggest that deficit irrigation accompanied with conservation tillage would greatly reduce the peanut crop water requirement.

For other crops, the application of deficit irrigation will result in a lower water use efficiency and yield. This is the case when crops are sensitive to drought stress throughout the complete season, such as maize.

Apart from university research groups and farmers associations, international organizations such as FAO, ICARDA, IWMI and the CGIAR Challenge Program on Water and Food are studying DI.

Reasons for Increased Water Productivity Under Deficit Irrigation

If crops have certain phenological phases in which they are tolerant to water stress, DI can increase the ratio of yield over crop water consumption (evapotranspiration) by either reducing the water loss by unproductive evaporation, and/or by increasing the proportion of marketable yield to the totally produced biomass (harvest index), and/or by increasing the proportion of total biomass production to transpiration due to hardening of the crop - although this effect is very limited due to the conservative relation between biomass production and crop transpiration, - and/or due to adequate fertilizer application and/or by avoiding bad agronomic conditions during crop growth, such as water logging in the root zone, pests and diseases, etc.

Advantages

The correct application of DI for a certain crop:

- Maximizes the productivity of water, generally with adequate harvest quality;

- Allows economic planning and stable income due to a stabilization of the harvest in comparison with rainfed cultivation;
- Decreases the risk of certain diseases linked to high humidity (e.g. fungi) in comparison with full irrigation;
- Reduces nutrient loss by leaching of the root zone, which results in better groundwater quality and lower fertilizer needs as for cultivation under full irrigation;
- Improves control over the sowing date and length of the growing period independent from the onset of the rainy season and therefore improves agricultural planning.

Constraints

A number of constraints apply to deficit irrigation:

- Exact knowledge of the crop response to water stress is imperative.
- There should be sufficient flexibility in access to water during periods of high demand (drought sensitive stages of a crop).
- A minimum quantity of water should be guaranteed for the crop, below which DI has no significant beneficial effect.
- An individual farmer should consider the benefit for the total water users community (extra land can be irrigated with the saved water), when he faces a below-maximum yield.
- Because irrigation is applied more efficiently, the risk for soil salinization is higher under DI as compared to full irrigation.

Modeling

Field experimentation is necessary for correct application of DI for a particular crop in a particular region. In addition, simulation of the soil water balance and related crop growth (crop water productivity modeling) can be a valuable decision support tool. By conjunctively simulating the effects of different influencing factors (climate, soil, management, crop characteristics) on crop production, models allow to (1) better understand the mechanism behind improved water use efficiency, to (2) schedule the necessary irrigation applications during the drought sensitive crop growth stages, considering the possible variability in climate, to (3) test DI strategies of specific crops in new regions, and to (4) investigate the effects of future climate scenarios or scenarios of altered management practices on crop production.

Flexible Barge

A flexible barge is a fabric barge (non-rigid) for the transportation of bulk fresh water or other liquid bulk items such as chemicals or oil.

Flexible barge air test.

History

One such barge is called the Dracone Barge invented in 1956. Other similar devices are the Spragg Bag invented in the 1980s, the water bag proposed by Nordic Water Supply in the late 1990s and the more recent REFRESH modular waterbag, developed in the 2010s. Terry Spragg of Manhattan Beach, California, builds flexible fabric barges for the transportation of bulk fresh water and is the reason why his product is referred to as the "Spragg Bag." In the 1970s Spragg was a promoter of icebergs as a large source of fresh water, but soon realized this was impractical. He then put his skills into developing the waterbag technology starting in the 1980s. Spragg has worked on and perfected this over the last twenty years with his associates. The first field test of his waterbag was in December 1990. The waterbag was 75 meters (246 feet) long and it contained approximately 3,000 cubic metres (790,000 US gal) of fresh water. It was towed from the Port Angeles harbor in the state of Washington. Another test was done in 1996 with a 100-mile (160 km) voyage from Port Angeles to Seattle, Washington. This test ended on April 29 when the fabric of one of the two bags under tow developed a tear. Spragg says that his next goal is to run another test voyage demonstration between Northern and Southern California and a demonstration of the waterbag technology in the Middle East as well as around the world. There are various reasons why it has been difficult to gain support for demonstrating the viability of waterbag technology in California and around the world. Spragg claims when two waterbags pass underneath the Golden Gate Bridge for the first time in history the media will let the whole world know about it. A novel, *Water, War, and Peace,* has been completed that details the solutions waterbag technology offers to the complex political problems surrounding water issues throughout the Middle East, the United States, and the world.

Nordic Water Supply waterbag.

The Norwegian company Nordic Water Supply developed a 10.800 m³ bag in 1997 under an agreement with the Turkish government to transport freshwater to Northern Cyprus. Within two years at least 7 million m³ of water had to be delivered annually at a cost of €2.7M per year, with volumes growing over time, but the actual transport only amounted to 4 million m³ in four years and the contract was discontinued by Turkish authorities. As a result, NWS went out of business and was de-listed from the Oslo Stock Exchange in 2003. NSW's waterbag technology was acquired by the Monohakobi Institute of Technology in Japan.

REFRESH waterbag.

The REFRESH waterbag was developed by a consortium of companies and research institutes from Greece, Spain, Italy, Turkey and the Czech Republic within two European FP7 projects, REFRESH (running from 2010 to 2012) and the follow-up XXL-REFRESH (running from 2013 to 2015). The first project was focused on validation of the concept of modular waterbag; it developed a small scale prototype of 200 m³ capacity, tested in Greece in 2012. The second project was focused on scale-up and partial redesign of the REFRESH system. At the end of the second project the REFRESH waterbag concept reached commercial scale and a 2500 m³ system made of five 500m³ modules was tested offshore Spain in 2015. The length of the waterbag was 60 m long.

The REFRESH concept is different from concepts of waterbag proposed earlier, based on huge monolithic containers (as the one proposed by Nordic Water Supply) or "trains" of smaller containers each one being sealed in itself (as the Spragg bag). The REFRESH waterbag is made of a series of modules, each one being a cylinder open at both bases, joined by watertight zippers. This makes it possible to perform all "dry" operations on ground at the level of single modules, overcoming the handling problems of monolithic containers and improves the behaviour in navigation with respect to the "trains" of connected bags.

Technology

Flexible barge filled with fresh water.

Flexible barge zipper.

The 1995 associated Spragg patents (#5,413,065 and #5,488,921) indicate that the inventions relate to a flexible fabric barge technology or combination of several barges made of a rubber polyurethane material. The main body portion of a flexible fabric barge is cylindrical in shape. The barge can be used by itself or as several connected flexible fabric barges that can be towed through the open ocean under extreme conditions. The patents further explain that the goal of Spragg's inventions are a practical water delivery system of fresh drinkable water that could be delivered to dry regions worldwide that have a shortage of potable water.

One of the flexible fabric barge concepts aims at an economical delivery system for fresh water that would be considerably cheaper than desalination plants, rigid ships, tanker trucks, conventional barges, aqueducts or pipeline transport. Waterbags are more economical and better for the environment than desalination of the seas and oceans.

The flexible fabric barge is a plastic container that is buoyant. It can be linked together with an attachment system to other flexible fabric barges to make a "train." The "waterbags", which are sometimes referred to as balloons or bladders, are made up of a vinyl lining inside a stronger material net. It is a type of "fabric water pipeline" when several are strung together for transport of liquids by tugboat through seas or open oceans to remote locations.

Zipper

Zippers play an important part in extending the capacity of the waterbag beyond what is practically achievable by a single textile piece. The Spragg and REFRESH concepts both feature prominently zippers, albeit with a fundamental difference in its function.

In the Spragg design, large waterbags are connected together like boxcars in a "train" fashion to increase the amount of liquids delivered at a time. It is estimated that the flexible barges could be as large as 14 meters in diameter and 150 meters in length, holding up to 17,000 cubic meters of fresh water or any other transportable liquid. Theoretically as many as 50 to 60 "waterbags" could be connected to one another and towed, although such a test has not been done to date.

REFRESH waterbag's fabric with embedded zipper.

Engineers suggest that the fabric barge can hold a "train" of 4,500,000 US gallons (17,000 m^3) of liquids.

A zipper coupler system is 10 times stronger than what is pulled by a 4,300 hp tug. Laborde Marine estimates a 4,300 horsepower tug with a bollard pull of 110,000 pounds can pull a "train" of fifty or so flexible barges weighing up to 1,300,000 tons. The "train" speed would be about 3 knots. This is over 700 acre feet (8,600,000 hl) of fresh water or other liquid per trip.

In the REFRESH design, the container itself is assembled on shore starting from planar cuts of fabric. Zippers run all along the perimeter of the fabric and make it possible to join an indefinite number of modules. Since each module is not closed by itself, zippers need to be watertight in order to ensure that no seawater leaks in.

Invention

Spragg Flexible barges coupled with zippers.

The basic invention is a device for the delivery of huge amounts of fresh water (700,000 to 4,500,000 US gallons (17,000 m^3) in each bag) at one time in a hostile wind and wave environment typical of oceans and large seas. Since freshwater is lighter than seawater the filled "bladders" (as they are sometimes referred to) float on top, similarly to icebergs, with little above the surface and most below the surface. Fresh water can be taken from rivers just before it discharges into salty seas or oceans, which then would not interfere with salmon spawning.

The associated coupler and zipper patent describes that to be economically feasible there should be several such flexible barges towed at one time. The greater the volume of water that can be delivered per trip, the better the economics. This string of barges would typically consist of barges in size from 25 to 50 feet (15 m) in diameter and 200 to 800 feet (240 m) in length each.

The unique characteristic of the Spragg Bag system is not the large volume of water in each bag, but what is called the world's strongest zipper (produced by Italian company Ziplast) that allows connection of several bags together in long trains. The large connecting zipper can be operated manually or by remote control with radio signals. The string of such flexible fabric "waterbags" may be coupled to a barge via a reinforced fabric nose cone where a tow line is attached.

The REFRESH scheme is enabled by a specialty zipper, again developed by Ziplast, that uses a completely different tooth engagement design able to keep the strength of the original "Spragg" zipper while adding watertightness. Tests performed by the Spanish research centre AIMPLAS have confirmed that the zipper is able to stay watertight even when in tension.

Each "waterbag" is generally filled to 80-90% capacity (so that it is not stiff and remains able to adapt to deformations when turning) and is flat across its top.

Economics

500,000 acre feet (620,000,000 m³) per year waterbag loading system designed by CH2M-HILL.

One San Francisco area reporter writes that waterbag technology would provide economical fresh water delivery to the Monterey Peninsula district and a solution to the shortage of fresh water in the area. He reports that the average family of four uses one acre-foot of water a year. This costs over $1000 for delivery using conventional methods, however this same amount of water delivered by Spragg Bags would cost about 30% less. Another newspaper reporter explains that towing Spragg Waterbags is environmentally friendly and is more economical than carrying water in ships or water tanker trucks or even using conventional rigid pipelines. An article in the July issue of ECONOMIST Magazine in 2008 explains that worldwide there is enough water for all, however most is often in the wrong place at the wrong time and it is just too expensive to transport.

Waterbag technology offers an easy and inexpensive solution to the problem of today's expensive conventional water transportation. It eliminates the difficulty of transporting water long distances by using the ocean as the means of transport. Waterbags considerably lower the capital costs and operating costs of moving fresh water from place to place. If a train was able to only move one or two box cars at a time, rather than in a train of dozens of boxcars at a time, it would not be very efficient and extremely expensive. Linking waterbags into "trains" of strings of waterbags and moving them through the ocean increases the economics of water transportation making it a viable practical option.

The cost to transport water 300 to 800 miles (1,300 km) through the ocean, based on deliveries of 5 million US gallons per day (19,000 m^3/d) to 10 million US gallons per day (38,000 m^3/d), is estimated to be between $350 to $450 per acre foot, depending on the length of the voyage and the amount of water delivered per trip. Increasing the amount of water delivered per day in each waterbag train will help to significantly reduce the cost of the water delivered. Once the reliability of the waterbag delivery system has proven its economics and reliability it will just be a matter of adding more waterbags to the trains, and more trains to the system in order to increase the amount of water delivered to selected locations, while also reducing the cost of the water delivered. Based on the increasing reliability of the waterbag delivery system over time, it should be possible to be able to economically deliver 100,000's of acre feet per year to many coastal locations around the world.

According to the inventor of the Spragg bag, the total cost of delivering fresh water down the California coast by his waterbag technology for a distance of 800 miles (1,300 km) from British Columbia to Monterey would cost about $966 per acre-foot per year. Keith Spain in a study for a Master Of Arts then shows in an analysis that it would save the residents of the Monterey Peninsula some $1,134 per-acre foot otherwise using a desalination plant. This is a savings of over $19 million per year for the Monterey taxpayers. This number assumes a usage of approximately 17,000 acre feet (21,000,000 m^3) per year (17,000 X $1,134 = $19,278,000 savings).

Applications

Flexible barge with camera crew on top filming an actual application.

One application seen is in the Middle East where large quantities of fresh water that are available in the Turkey region could be delivered to other places around the Mediterranean Sea that have an extreme shortage of drinkable fresh water, like Israel and Gaza. Spragg believes that delivering fresh drinking water to water-poor nations can promote world peace.

Israeli President Shimon Peres has written a letter in support of implementing a demonstration of Spragg Bag technology in the Mediterranean Sea as a tool for helping to bring Peace to the Middle East. In this letter President Peres states, "The draft of WATER, WAR AND PEACE written as a novel is in my view an original approach to highlight this grave problem and its solutions, that will pave the path to a better and more peaceful region. Your efforts to embark on a demonstration voyage to enlighten us all, both regarding the technological viability as well as cost, will surely contribute to meet the critical dilemma."

He also points out that using waterbags towed through the Mediterranean Sea would be much more economical than transporting water through pipeline systems. This view is shared by the REFRESH consortium.

Waterbags have been proposed to be used for emergency purposes in order to link the Gulf Cooperation Counsel countries' desalination plants all along the Persian Gulf coast.

Another application is the regular delivery of fresh water over long distance routes.

Proposed routes include from the state of Washington to dry regions of Southern California, from Mad River in Northern California to the San Francisco area and from Southern Chile to Atacama. Other applications are that it:

- Could be used to move water through the Sacramento River Delta following an earthquake and a catastrophic levee collapse that could cut off Southern California's water supply for up to two years or more.
- Could deliver large quantities of stormwater and/or recycled water to areas that need more fresh water to offset lower water levels and rising salinity.
- Could be pre-positioned storage worldwide of large quantities of fresh drinking water for quick delivery after a natural disaster.

Spragg has proposed to deliver water from the Manavgat River in Turkey across the Mediterranean Sea to Israel and the Gaza Strip, which has an extreme shortage of water, which presently is being reviewed by the World Bank.

Spragg has proposed to the Australian government that bulk fresh drinking water using his waterbag technology could be applied to urban water supplies that have a shortage. It would establish an economically sound new industry for bulk fresh water transport. An analysis of the economic and environmental advantages for waterbag technology by using the ocean currents from the Tully River to Brisbane, Queensland has been completed. On the Gold Coast the Tugan desalination plant is being built to supply 120 megaliters of drinking water daily at a cost of approximately $1.2 billion. Operation of the Tugan desalination plant will produce 235,000 tons of CO_2 greenhouse gases annually. Preliminary cost estimates indicate that using waterbag technology to deliver the same amount of water from the Tully river to the Gold Coast may be 30 times less expensive than desalination, and that waterbag technology may emit 60 times less greenhouse gas. These figures suggest that waterbag technology would deliver water for much less cost than the proposed pipeline from the Burdekin River to Brisbane which is projected to cost approximately $7.5 billion for infrastructure and about $250 million for annual operation.

Spragg has proposed to the White House the idea of a peace offering in the Middle East by supplying 20 to 30 of his waterbags with fresh water and transport them from Turkey to the Palestinians and Israelis. The technology could also be used for the collection of Australia's factory waste water outputs, storm water, and sewerage for processing and reuse.

Hippo Water Roller

Upper: traditional method of carrying water; lower: more water can be transported easily using the roller.

The Hippo water roller, or Hippo roller, is a device for carrying water more easily and efficiently than traditional methods, particularly in the developing world. It consists of a barrel-shaped container which holds the water and can roll along the ground, and a handle attached to the axis of the barrel. Currently deployed in rural Africa, its simple and purpose-built nature makes it an example of appropriate technology.

Physical Design

The drum of the Hippo water roller is made from UV stabilized linear low-density polyethylene and is designed to cope with the rough surfaces found in rural areas. The drum's volume is 90 liters (about 24 gallons). It has a large opening (135 mm / 5.3 inch diameter) for easy filling and cleaning. The size of the opening was originally determined by the availability of a large enough lid. The lid then on its part determined the roll radius of the roller, since enough clearance was required with the lid fitted to keep the latter clear from the ground and any obstacles which might damage it. The lid was eventually further recessed on later models to enhance its protection.

The steel handle allows the roller to be pushed or pulled over difficult and very rough terrain. The overall width of the roller with handle attached was determined by measuring the average width of a standard doorway and sized to allow it to be pulled through freely. This parameter together with the roll radius and the rounded shoulders eventually determined the average volume of 90 liters. The steel handle is fitted with special polymer end-caps to reduce friction and wear and prolong the life of the pivot cavities in the drum.

During development a water filled roller was drawn behind a vehicle over a dirt road at 20 km/h for 15 km without any significant signs of wear on both the roller outer surface or pivot cavities.

The roller is rounded at the shoulders to simplify tilting when wanting to pour from the full roller. However, the roller is also very stable in the upright position when it rests on a small, flat surface. The roller has hand grips at the bottom and top to make emptying the container easier.

History

The barrel, originally called "*Aqua Roller*", was the brainchild of two engineers, Pettie Petzer and Johan Jonker of South Africa

Petzer and Jonker were recognized for their work on the Hippo Roller in 1997 with the "Design for Development Award" by the South African Bureau of Standards and its Design Institute.

Claimed Benefits

It is claimed that approximately five times the amount of water can be transported in less time with far less effort than the traditional method of carrying 20 liters (approximately 5 gallons) on the head.

Claimed benefits include:

- Time savings (fetching water can be very time consuming in some poor rural environments);
- Reduced effort;
- Reduced strain (carrying heavy weights on the head every day for years puts strain on the body, particularly the vertebral column);
- Increased water availability, with benefits for health and perhaps even enabling vegetables to be grown;
- Hygienic storage due to the sealed lid on the roller.

Use

Although the Hippo roller was mainly designed as a water gathering aid for underprivileged communities, several other uses have since been discovered. Because of the Hippo water roller's water tightness it can act as a watertight container to keep valuables and clothes dry when going downstream in white water rapids on extreme excursions.

It can also suffice as a container to be dropped with life saving contents from helicopters or low flying fixed-wing aircraft in flooded disaster areas. This might include food and medical supplies, warm dry clothing and blankets, and naturally, water. The polyethylene is lighter than water and will stay only partly submerged, even with fresh water in it. In fact, some rollers have been distributed pre-filled with grain or other foodstuff to villages.

One of the draw-backs of the distribution of the rollers is that the rollers cannot be stacked efficiently to save space; thus transport capacity is wasted.

Administration

Infotech, an information technology company, initially sponsored the project as a social responsibility project known as the *Hippo Water Roller Trust Fund.* Currently, the project is supported financially by donor funding which comes from individuals, corporate businesses and non-profit organization partners. [Hippo Water International] is a non-governmental organization in the United States created to raise funds there.

Deployment

46,000 Hippo rollers have been manufactured and distributed so far.

As part of the investigation into the alleviation of poverty and scarcity of water in the far northern parts of Namibia, the Social Sciences Division of the Multi-Disciplinary Research Centre at the University of Namibia, sponsored by UNICEF's Directorate of Rural Development bought 1000 rollers and distributed them into the community. Six months after the introduction they launched an evaluation on the success and performance of the rollers as well as the social impact of the roller on the lives of the recipients.

Peak Water

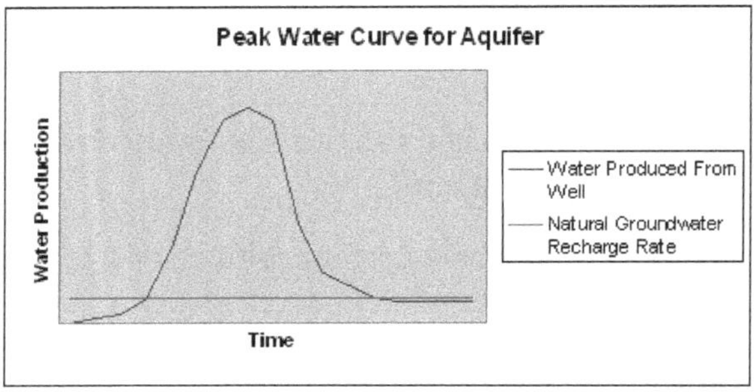

Potential peak water curve for production of groundwater from an aquifer.

Peak water is a concept that underlines the growing constraints on the availability, quality, and use of freshwater resources.

Peak water is defined in a 2010 peer-reviewed article in the *Proceedings of the National Academy of Sciences* by Peter Gleick and Meena Palaniappan. They distinguish between peak renewable, peak non-renewable, and peak ecological water in order to demonstrate the fact that although there is a vast amount of water on the planet, sustainably managed water is becoming scarce.

Lester R. Brown, president of the Earth Policy Institute, wrote in 2013 that although there was extensive literature on peak oil, it was peak water that is "the real threat to our future". An assessment was published in August 2011 in the Stockholm International Water Institute's journal. Much of the world's water in underground aquifers and in lakes can be depleted and thus resembles a finite resource. The phrase *peak water* sparks debates similar to those about peak oil. In 2010, *New York Times* chose "peak water" as one of its 33 "Words of the Year".

There are concerns about impending peak water in several areas around the world:

- Peak renewable water, where entire renewable flows are being consumed for human use.
- Peak non-renewable water, where groundwater aquifers are being overpumped (or contaminated) faster than nature recharges them (this example is most like the peak oil debate).
- Peak ecological water, where ecological and environmental constraints are overwhelming the economic benefits provided by water use.

If present trends continue, 1.8 billion people will be living with absolute water scarcity by 2025, and two-thirds of the world could be subject to water stress. Ultimately, peak water is not about running out of freshwater, but about reaching physical, economic, and environmental limits on meeting human demands for water and the subsequent decline of water availability and use.

Comparison with Peak Oil

The Hubbert curve has become popular in the scientific community for predicting the depletion of various natural resources. M. King Hubbert created this measurement device in 1956 for a variety of finite resources such as coal, oil, natural gas and uranium. Hubbert's curve was not applied to resources such as water originally, since water is a renewable resource. Some forms of water, however, such as fossil water, exhibit similar characteristics to oil, and overpumping (faster than the rate of natural recharge of groundwater) often results in a Hubbert-type peak. A modified Hubbert curve applies to any resource that can be harvested faster than it can be replaced. Like peak oil, peak water is inevitable given the rate of extraction of certain water systems. A current argument is that growing populations and demands for water will inevitably lead to non-renewable use of water resources.

Water Supply

Fresh water is a renewable resource, yet the world's supply of clean, fresh water is under increasing demand for human activities. The world has an estimated 1.34 billion cubic kilometers of water, but 96.5% of it is salty. Almost 70% of fresh water can be found in the ice caps of Antarctica and Greenland. Less than 1% of this water on Earth is accessible to humans, the rest is contained in soil moisture or deep underground. Accessible freshwater is located in lakes, rivers, reservoirs and shallow underground sources. Rainwater and snowfall do very little to replenish these stocks of freshwater.

Freshwater sources (top 15 countries)		
Country	**Total freshwater supply**	
	(km^3/yr)	**Year**
Brazil	8233	2000
Russia	4498	1997
Canada	3300	1985
Colombia	3132	2000

USA	3069	1985
Indonesia	2838	1999
China	2830	2008
Peru	1913	2000
India	1908	1999
DR Congo	1283	2001
Venezuela	1233	2000
Bangladesh	1211	1999
Burma	1046	1999
Chile	922	2000
Vietnam	891	1999

The amount of available freshwater supply in some regions is decreasing because of (i) climate change, which has caused receding glaciers, reduced stream and river flow, and shrinking lakes; (ii) contamination of water by human and industrial wastes; and (iii) overuse of non-renewable groundwater aquifers. Many aquifers have been over-pumped and are not recharging quickly. Although the total freshwater supply is not used up, much has become polluted, salted, unsuitable or otherwise unavailable for drinking, industry and agriculture.

Water Demand

Water demand already exceeds supply in many parts of the world, and as the world population continues to rise, many more areas are expected to experience this imbalance in the near future.

Agriculture represents 70% of freshwater use worldwide.

Agriculture, industrialization and urbanization all serve to increase water consumption.

Freshwater Withdrawal by Country

The highest total annual consumption of water comes from India, China and the United States, countries with large populations, extensive agricultural irrigation, and demand for food. See the following table:

Freshwater withdrawal by country and sector (top 20 countries)					
Country	**Total freshwater withdrawal (km^3/yr)**	**Per capita withdrawal (m^3/p/yr)**	**Domestic use (m^3/p/yr) (in %)**	**Industrial use (m^3/p/yr)(in %)**	**Agricultural use (m^3/p/yr) (in %)**
India	645.84	585	47 (8%)	30 (5%)	508 (86%)
China	549.76	415	29 (7%)	107 (26%)	279 (68%)
United States	477	1,600	208 (13%)	736 (46%)	656 (41%)

Freshwater withdrawal by country and sector (top 20 countries)					
Country	Total freshwater withdrawal (km^3/yr)	Per capita withdrawal ($m^3/p/yr$)	Domestic use ($m^3/p/yr$) (in %)	Industrial use ($m^3/p/yr$)(in %)	Agricultural use ($m^3/p/yr$) (in %)
Pakistan	169.39	1,072	21 (2%)	21 (2%)	1029 (96%)
Japan	88.43	690	138 (20%)	124 (18%)	428 (62%)
Indonesia	82.78	372	30 (8%)	4 (1%)	339 (91%)
Thailand	82.75	1,288	26 (2%)	26 (2%)	1236 (95%)
Bangladesh	79.4	560	17 (3%)	6 (1%)	536 (96%)
Mexico	78.22	731	126 (17%)	37 (5%)	569 (77%)
Russia	76.68	535	102 (19%)	337 (63%)	96 (18%)
Iran	72.88	1,048	73 (7%)	21 (2%)	954 (91%)
Vietnam	71.39	847	68 (8%)	203 (24%)	576 (68%)
Egypt	68.3	923	74 (8%)	55 (6%)	794 (86%)
Brazil	59.3	318	64 (20%)	57 (18%)	197 (62%)
Uzbekistan	58.34	2,194	110 (5%)	44 (2%)	2040 (93%)
Canada	44.72	1,386	274 (20%)	947 (69%)	165 (12%)
Iraq	42.7	1,482	44 (3%)	74 (5%)	1363 (92%)
Italy	41.98	723	130 (18%)	268 (37%)	325 (45%)
Turkey	39.78	544	82 (15%)	60 (11%)	403 (74%)
Germany	38.01	460	55 (12%)	313 (68%)	92 (20%)

India

Working rice paddies.

India has 20 percent of the Earth's population, but only four percent of its water. Water tables are dropping rapidly in some of India's main agricultural areas. The Indus and Ganges rivers are tapped so heavily that, except in rare wet years, they no longer reach the sea.

India has the largest water withdrawal out of all the countries in the world. Eighty-six percent of that water supports agriculture. That heavy use is dictated in large part by what people eat. People in India consume a lot of rice. Rice farmers in India typically get less than half the yield per unit area while using ten times more water than their Chinese counterparts. Economic development can make things worse because as people's living standards rise, they tend to eat more meat, which requires lots of water to produce. Growing a tonne of grain requires 1,000 tonnes of water; producing a tonne of beef requires 15,000 tonnes. To make a single hamburger requires around 4,940 liters (1,300 gallons) of water A glass of orange juice needs 850 liters (225 gallons) of freshwater to produce.

China

China, the world's most populous country, has the second largest water withdrawal; 68% supports agriculture while its growing industrial base consumes 26%. China is facing a water crisis where water resources are overallocated, used inefficiently, and severely polluted by human and industrial wastes. One-third of China's population lacks access to safe drinking water. Rivers and lakes are dead and dying, groundwater aquifers are over-pumped, uncounted species of aquatic life have been driven to extinction, and direct adverse impacts on both human and ecosystem health are widespread and growing.

In western China's Qinghai province, through which the Yellow River's main stream flows, more than 2,000 lakes have disappeared over the last 20 years. There were once 4,077 lakes. Global climate change is responsible for the reduction in flow of the (Huang He) Yellow River over the past several decades. The source of the Yellow River is the Qinghai-Xizang Tibetan Plateau where the glaciers are receding sharply.

In Hebei Province, which surrounds Beijing, the situation is much worse. Hebei is one of China's major wheat and corn growing provinces. The water tables have been falling fast throughout Hebei. The region has lost 969 of its 1,052 lakes. About 500,000 people are affected by a shortage of drinking water due to continuing droughts. Hydro-power generation is also impacted. Beijing and Tianjin depend on Hebei Province to supply their water from the Yangtze River. Beijing gets its water via the newly constructed South-North Water Transfer Project. The river originates in a glacier on the eastern part of the Tibetan Plateau.

United States

Ship canal terminus.

The United States has about 5% of the world's population, yet it uses almost as much water as India (~1/5 of world) or China (1/5 of world) because substantial amounts of water are used to grow food exported to the rest of the world. The U.S. agricultural sector consumes more water than the industrial sector, though substantial quantities of water are withdrawn (but not consumed) for power plant cooling systems. 40 out of 50 state water managers expect some degree of water stress in their state in the next 10 years.

The Ogallala Aquifer in the southern high plains (Texas and New Mexico) is being mined at a rate that far exceeds replenishment—a classic example of peak non-renewable water. Portions of the aquifer will not naturally recharge due to layers of clay between the surface and the water-bearing formation, and because rainfall rates simply do not match rates of extraction for irrigation. The term fossil water is sometimes used to describe water in aquifers that was stored over centuries to millennia. Use of this water is not sustainable when the recharge rate is slower than the rate of groundwater extraction.

In California, massive amounts of groundwater are also being withdrawn from Central Valley groundwater aquifers—unreported, unmonitored, and unregulated. California's Central Valley is home to one-sixth of all irrigated land in the United States, and the state leads the nation in agricultural production and exports. The inability to sustain groundwater withdrawals over time may lead to adverse impacts on the region's agricultural productivity.

The Central Arizona Project (CAP) is a 336-mile (541 km) long canal that diverts 489 billion US gallons (1.85×10^9 m^3) a year from the Colorado River to irrigate more than 300,000 acres (1,200 km^2) of farmland. The CAP project also provides drinking water for Phoenix and Tucson. It has been estimated that Lake Mead, which dams the Colorado, has a 50-50 chance of running dry by 2021.

The Ipswich River near Boston now runs dry in some years due to heavy pumping of groundwater for irrigation. Maryland, Virginia and the District of Columbia have been fighting over the Potomac River. In drought years like 1999 or 2003, and on hot summer days the region consumes up to 85 percent of the river's flow.

Per Capita Withdrawal of Water

Turkmenistan, Kazakhstan and Uzbekistan use the most water per capita. See the table below:

Freshwater withdrawal by country and sector (top 15 countries, per capita)					
Country	Total freshwater withdrawal	Per capita withdrawal	Domestic use	Industrial use	Agricultural use
	(km³/yr)	(m³/p/yr)	(%)	(%)	(%)
urkmenistan	24.65	5,104	2	1	98
Kazakhstan	35	2,360	2	17	82
Uzbekistan	58.34	2,194	5	2	93
Guyana	1.64	2,187	2	1	98
Hungary	21.03	2,082	9	59	32
Azerbaijan	17.25	2,051	5	28	68
Kyrgyzstan	10.08	1,916	3	3	94
Tajikistan	11.96	1,837	4	5	92

Freshwater withdrawal by country and sector (top 15 countries, per capita)					
Country	Total freshwater withdrawal	Per capita withdrawal	Domestic use	Industrial use	Agricultural use
	(km³/yr)	(m³/p/yr)	(%)	(%)	(%)
USA	477	1,600	13	46	41
Suriname	0.67	1,489	4	3	93
Iraq	42.7	1,482	3	5	92
Canada	44.72	1,386	20	69	12
Thailand	82.75	1,288	2	2	95
Ecuador	16.98	1,283	12	5	82
Australia	24.06	1,193	15	10	75

Turkmenistan

Orphaned ship in former Aral Sea, near Aral, Kazakhstan.

Turkmenistan gets most of its water from the Amu Darya River. The Qaraqum Canal is a canal system that takes water from the Amu Darya River and distributes the water out over the desert for irrigation of its orchard crops and cotton. Turkmenistan uses the most water per capita in the world because only 55% of the water delivered to the fields actually reaches the crops.

Kazakhstan and Uzbekistan

The two rivers feeding the Aral Sea were dammed up and the water was diverted to irrigate the desert so that cotton could be produced. As a result, the Aral Sea's water has become much saltier and the sea's water level has decreased by over 60%. Drinking water is now contaminated with pesticides and other agricultural chemicals and contains bacteria and viruses. The climate has become more extreme in the area surrounding it.

Water Shortfall by Country

Saudi Arabia, Libya, Yemen and United Arab Emirates have hit peaks in water production and are depleting their water supply. See the table below:

Freshwater shortfall by country (top 15 countries)			
Region and country	Total freshwater withdrawal (km³/yr)	Total freshwater supply (km³/yr)	Total freshwater shortfall (km³/yr)
Saudi Arabia	17.32	2.4	14.9
Libya	4.27	0.6	3.7
Yemen	6.63	4.1	2.5
United Arab Emirates	2.3	0.2	2.2
Kuwait	0.44	0.02	0.4
Oman	1.36	1.0	0.4
Israel	2.05	1.7	0.4
Qatar	0.29	0.1	0.2
Bahrain	0.3	0.1	0.2
Jordan	1.01	0.9	0.1
Barbados	0.09	0.1	0.0
Maldives	0.003	0.03	0.0
Antigua and Barbuda	0.005	0.1	0.0
Malta	0.02	0.07	-0.1
Cyprus	0.21	0.4	-0.2

Saudi Arabia

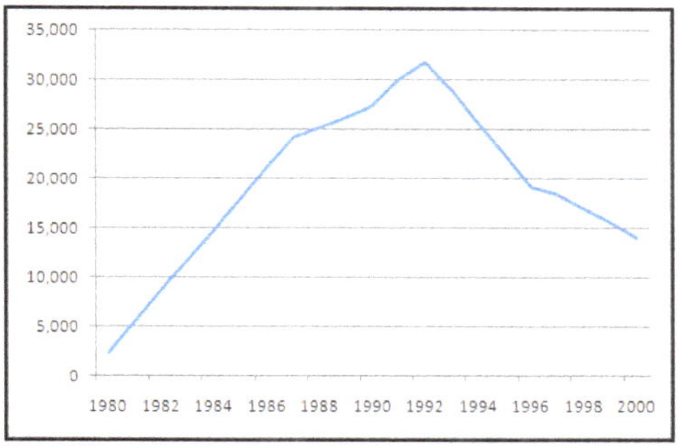

Water supply in Saudi Arabia, 1980–2000, in millions of cubic meters.

According to Walid A. Abderrahman (2001), "Water Demand Management in Saudi Arabia", Saudi Arabia reached peak water in the early 1990s, at more than 30 billion cubic meters per year, and declined afterward. The peak had arrived at about midpoint, as expected for a Hubbert curve. Today, the water production is about half the peak rate. Saudi Arabian food production has been based on "fossil water"—water from ancient aquifers that is being recharged very slowly, if at all. Like oil, fossil water is non-renewable, and it is bound to run out someday. Saudi Arabia has abandoned its self-sufficient food production and is now importing virtually all of its food. Saudi Arabia has built desalination plants to provide about half the country's freshwater. The remainder comes from groundwater (40%), surface water (9%) and reclaimed wastewater (1%).

Libya

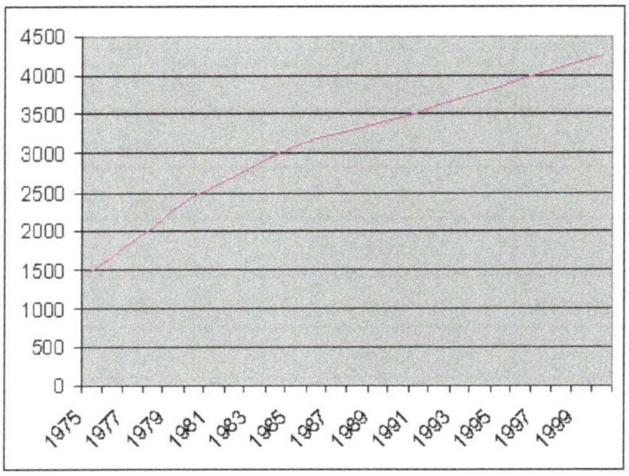

Water supply in Libya, 1975–2000, in millions of cubic meters.

Libya is working on a network of water pipelines to import water, called the Great Manmade River. It carries water from wells tapping fossil water in the Sahara desert to the cities of Tripoli, Benghazi, Sirte and others. Their water also comes from desalination plants.

Yemen

Peak water has occurred in Yemen. Sustainability is no longer attainable in Yemen, according to the government's five-year water plan for 2005–2009. The aquifer that supplies Sana'a, the capital of Yemen, will be depleted by 2009. In its search for water in the basin, the Yemeni government has drilled test wells that are 2 kilometers (1.2 mi) deep, depths normally associated with the oil industry, but it has failed to find water. Yemen must soon choose between relocating the city and building a pipeline to coastal desalination plants. The pipeline option is complicated by Sana'a's altitude of 2,250 m (7,380 ft).

As of 2010, the threat of running out of water was considered greater than that of Al-Qaeda or instability. There was speculation that Yemenis would have to abandon mountain cities, including Sana'a, and move to the coast. The cultivation of khat and poor water regulation by the government were partly blamed.

United Arab Emirates

Desalination plant in Ras al-Khaimah, UAE.

United Arab Emirates has a rapidly growing economy and very little water to support it. UAE requires more water than is naturally available. They have reached peak water. To solve this, UAE has a desalination plant near Ruwais and ships its water via pipeline to Abu Dhabi.

Consequences

Famine

Water shortage may cause famine in Pakistan. Pakistan has approximately 35 million acres (140,000 km^2) of arable land irrigated by canals and tube wells, mostly using water from the Indus River. Dams were constructed at Chashma, Mangla, and Tarbela to feed the irrigation system. Since the completion of the Tarbela Dam in 1976 no new capacity has been added despite astronomical growth in population. The gross capacity of the three dams has decreased because of sedimentation, a continual process. Per-capita surface-water availability for irrigation was 5,260 cubic meters per year in 1951. This has been reduced to a mere 1,100 cubic meters per year in 2006.

Health Problems

The quality of drinking water is vital for human health. Peak water constraints result in people not having access to safe water for basic personal hygiene. "Infectious waterborne diseases such as diarrhea, typhoid, and cholera are responsible for 80 percent of illnesses and deaths in the developing world, many of them children. One child dies every eight seconds from a waterborne disease; 15 million children a year."

Vital aquifers everywhere are becoming contaminated with toxins. Once an aquifer is contaminated, it is not likely that it can ever recover. Contaminants are more likely to cause chronic health effects. Water can be contaminated from pathogens such as bacteria, viruses, and parasites. Also, toxic organic chemicals can be a source of water contamination. Inorganic contaminants include toxic metals like arsenic, barium, chromium, lead, mercury, and silver. Nitrates are another source of inorganic contamination. Finally, leaching radioactive elements into the water supply can contaminate it.

Human Conflicts Over Water

Some conflicts of the future may be fought over the availability, quality, and control of water. Water has also been used as a tool in conflicts or as a target during conflicts that start for other reasons. Water shortages may well result in water conflicts over this precious resource.

In West Africa and other places like Nepal, Bangladesh, India (such as the Ganges Delta), and Peru, major changes in the rivers generate a significant risk of violent conflict in coming years. Water management and control could play a part in future resource wars over scarce resources.

Solutions

Freshwater usage has great potential for better conservation and management as it is used inefficiently nearly everywhere, but until actual scarcity hits, people tend to take access to freshwater for granted.

Water Conservation

There are several ways to reduce the use of water. For example, most irrigation systems waste water; typically, only between 35% and 50% of water withdrawn for irrigated agriculture ever reaches the crops. Most soaks into unlined canals, leaks out of pipes, or evaporates before reaching (or after being applied to) the fields. Swales and cisterns can be used to catch and store excess rainwater.

Water should be used more efficiently in industry, which should use a closed water cycle if possible. Also, industry should prevent polluting water so that it can be returned into the water cycle. Whenever possible, gray wastewater should be used to irrigate trees or lawns. Water drawn from aquifers should be recharged by treating the wastewater and returned to the aquifer.

Water can be conserved by not allowing freshwater to be used to irrigate luxuries such as golf courses. Luxury goods should not be produced in areas where freshwater has been depleted. For example, 1,500 liters of water is used on average for the manufacturing of a single computer and monitor.

Water Management

Sustainable water management involves the scientific planning, developing, distribution and optimization of water resources under defined water polices and regulations. Examples of policies that improve water management include the use of technology for efficiency monitoring and use of water, innovative water prices and markets, irrigation efficiency techniques, and much more.

Experience shows that higher water prices lead to improvements in the efficiency of use—a classical argument in economics, pricing, and markets. For example, Clark County, Nevada, raised its water rates in 2008 to encourage conservation. Economists propose to encourage conservation by adopting a system of progressive pricing whereby the price per unit of water used would start out very small, and then rise substantially for each additional unit of water used. This tiered-rate approach has been used for many years in many places, and is becoming more widespread. A Freakonomics column in the *New York Times* similarly suggested that people would respond to higher water prices by using less of it, just as they respond to higher gasoline prices by using less of it. The *Christian Science Monitor* has also reported on arguments that higher water prices curb waste and consumption.

Conversely, certain subsidies can lead to inefficient use of water. Water subsidies often involve contentious policy issues that are political in nature. In 2004, the Environmental Working Group criticized the U.S. government for selling subsidized water to corporate farms for an average price of only $17 per acre foot (326,000 gallons).

In his book *The Ultimate Resource 2*, Julian Simon claimed that there is a strong correlation between government corruption and lack of sufficient supplies of safe, clean water. Simon wrote, "there is complete agreement among water economists that all it takes to ensure an adequate supply for agriculture as well as for households in rich countries is that there be a rational structure of water law and market pricing. The problem is not too many people but rather defective laws and bureaucratic interventions; freeing up markets in water would eliminate just about all water problems forever... In poor water-short countries the problem with water supply—as with so many other matters—is lack of wealth to create systems to supply water efficiently enough. As these countries become richer, their water problems will become less difficult". This theoretical argument, however, ignores real-world conditions, including strong barriers to open water markets, the difficulty of moving water from one region to another, water rights laws that prevent redistribution based solely on economic value, inability of some populations to pay for water, and grossly imperfect information on water use. Actual experience with peak water constraints in some wealthy, but water-short countries and regions still suggests serious difficulties in reducing water challenges.

Climate Change

Extensive research has shown the direct links between water resources, the hydrologic cycle, and climatic change. As climate changes, there will be substantial impacts on water demands, precipitation patterns, storm frequency and intensity, snowfall and snowmelt dynamics, and more. Evidence from the IPCC to Working Group II, has shown climate change is already having a direct effect on animals, plants and water resources and systems. A 2007 report by the Intergovernmental Panel on Climate Change counted 75 million to 250 million people across Africa who could face water shortages by 2020. Crop yields could increase by 20% in East and Southeast Asia, but decrease by up to 30% in Central and South Asia. Agriculture fed by rainfall could drop by 50% in some African countries by 2020. A wide range of other impacts could affect peak water constraints.

Loss of biodiversity can be attributed largely to the appropriation of land for agroforestry and the effects of climate change. The 2008 IUCN Red List warns that long-term droughts and extreme weather puts additional stress on key habitats and, for example, lists 1,226 bird species as threatened with extinction, which is one-in-eight of all bird species.

Backstop Water Sources

The concept of a "backstop" resource is a resource that is sufficiently abundant and sustainable to replace non-renewable resources. Thus, solar and other renewable energy sources are considered "backstop" energy options for unsustainable fossil fuels. Similarly, Gleick and Palaniappan defined "backstop water sources" to be those resources that can replace unsustainable and non-renewable use of water, albeit typically at a higher cost. The classic backstop water source is desalination of seawater. If the rate of water production is not sufficient in one area, another "backstop" could be increased interbasin transfers, such as pipelines to carry freshwater from where it is abundant to an area where water is needed. Water can be imported into an area using water trucks. The most expensive and last resort measures of getting water to a community such as desalination, water transfers are called "backstop" water sources. Fog catchers are the most extreme of backstop methods.

To produce that fresh water, it can be obtained from ocean water through desalination. A January 17, 2008 article in the *Wall Street Journal* stated, "World-wide, 13,080 desalination plants produce more than 12 billion US gallons (45,000,000 m^3) of water a day, according to the International Desalination Association". Israel is now desalinizing water at a cost of US$0.53 per cubic meter. Singapore is desalinizing water for US$0.49 per cubic meter. After being desalinized at Jubail, Saudi Arabia, water is pumped 200 miles (320 km) inland though a pipeline to the capital city of Riyadh.

However, several factors prevent desalination from being a panacea for water shortages:

- High capital costs to build the desalination plant.

- High cost of the water produced.
- Energy required to desalinate the water.
- Environmental issues with the disposal of the resulting brine.
- High cost of transporting water.

Nevertheless, some countries like Spain are increasingly relying on desalination because of the continuing decreasing costs of the technology.

At last resort, it is possible in some particular regions to harvest water from fog using nets. The water from the nets drips into a tube. The tubes from several nets lead to a holding tank. Using this method, small communities on the edge of deserts can get water for drinking, gardening, showering and clothes washing. Critics say that fog catchers work in theory but have not succeeded as well in practice. This is due to the high expense of the nets and pipe, high maintenance costs and low quality of water.

An alternative approach is that of the Seawater Greenhouse, which consists in desalinating seawater through evaporation and condensation inside a greenhouse solely using solar energy. Successful pilots have been conducted growing crops in desert locations.

References

- H. Cooley; P.H. Gleick; G. Wolff (2006). "Desalination, With a Grain of Salt". Pacific Institute, Oakland, California. Retrieved 01, March 2020
- "Development of Modular reconfigurable waterbag for long haul freshwater transport over sea, presentation given at Malta Water Week 2015"
- Van Ginkel, J. A. (2002). Human Development and the Environment: Challenges for the United Nations in the New Millennium. United Nations University Press. pp. 198–199. ISBN 9280810693

Permissions

All chapters in this book are published with permission under the Creative Commons Attribution Share Alike License or equivalent. Every chapter published in this book has been scrutinized by our experts. Their significance has been extensively debated. The topics covered herein carry significant information for a comprehensive understanding. They may even be implemented as practical applications or may be referred to as a beginning point for further studies.

We would like to thank the editorial team for lending their expertise to make the book truly unique. They have played a crucial role in the development of this book. Without their invaluable contributions this book wouldn't have been possible. They have made vital efforts to compile up to date information on the varied aspects of this subject to make this book a valuable addition to the collection of many professionals and students.

This book was conceptualized with the vision of imparting up-to-date and integrated information in this field. To ensure the same, a matchless editorial board was set up. Every individual on the board went through rigorous rounds of assessment to prove their worth. After which they invested a large part of their time researching and compiling the most relevant data for our readers.

The editorial board has been involved in producing this book since its inception. They have spent rigorous hours researching and exploring the diverse topics which have resulted in the successful publishing of this book. They have passed on their knowledge of decades through this book. To expedite this challenging task, the publisher supported the team at every step. A small team of assistant editors was also appointed to further simplify the editing procedure and attain best results for the readers.

Apart from the editorial board, the designing team has also invested a significant amount of their time in understanding the subject and creating the most relevant covers. They scrutinized every image to scout for the most suitable representation of the subject and create an appropriate cover for the book.

The publishing team has been an ardent support to the editorial, designing and production team. Their endless efforts to recruit the best for this project, has resulted in the accomplishment of this book. They are a veteran in the field of academics and their pool of knowledge is as vast as their experience in printing. Their expertise and guidance has proved useful at every step. Their uncompromising quality standards have made this book an exceptional effort. Their encouragement from time to time has been an inspiration for everyone.

The publisher and the editorial board hope that this book will prove to be a valuable piece of knowledge for students, practitioners and scholars across the globe.

Index

CPSIA information can be obtained
at www.ICGtesting.com
Printed in the USA
LVHW060826120122
708212LV00006B/414

9 781639 874590